AF614733

Unveiling the Mycorrhizal World

Edited by Everlon Cid Rigobelo

Published in London, United Kingdom

Unveiling the Mycorrhizal World
http://dx.doi.org/10.5772/intechopen.114574
Edited by Everlon Cid Rigobelo

Contributors
Ban Taha Mohammed, Everlon Cid Rigobelo, Fang Liu, Faqiang Zhan, Hongmei Yang, Huifang Bao, Kai Lou, Ming Chu, Ning Wang, Qing Lin, Rong Yang, Sana Jabeen, XinXiang Niu, Xuanqi Long, YingWu Shi, Yunjian Xu

First published in London, United Kingdom, 2024 by IntechOpen
IntechOpen is the global imprint of INTECHOPEN LIMITED, registered in England and Wales, registration number: 11086078, 167-169 Great Portland Street, London, W1W 5PF, United Kingdom

British Library Cataloguing-in-Publication Data
A catalogue record for this book is available from the British Library

Additional hard and PDF copies can be obtained from orders@intechopen.com

Unveiling the Mycorrhizal World
Edited by Everlon Cid Rigobelo
p. cm.
Print ISBN 978-0-85014-348-5
Online ISBN 978-0-85014-349-2
eBook (PDF) ISBN 978-0-85014-350-8

Meet the editor

Dr. Everlon Cid Rigobelo graduated from the Agronomy School Universidade Estadual Paulista Campus of Jaboticabal, Brazil, in 2000. He received his MSc in Agricultural Microbiology and Ph.D. in Microbiology from the same university. Dr. Rigobelo has experience in genetics and epidemiology and is active in microbial biotechnology, molecular genetics, bacterial genomics, and plant growth-promoting microorganisms. He works with plant growth-promoting rhizobacteria and has published more than 100 related studies. He has also edited several books. Nowadays, Dr. Rigobelo teaches microbiology and works toward reducing the dependency on fertilizers and pesticides in plant production. He aims to use microorganisms for sustainable production, thus reducing adverse environmental impacts and production costs.

Contents

Preface

Fungi are eukaryotic microorganisms that have structures called hyphae, which are similar to plant roots. These hyphae absorb water and nutrients from the soil and provide support to the fungi. Unicellular fungi that do not produce hyphae are referred to as yeasts. Fungi exhibit a vast range of metabolic and physiological capabilities, allowing them to produce an extensive array of secondary metabolites as well as a variety of enzymes that play critical roles in the soil's biochemical cycles. Additionally, these organisms play a crucial role in increasing the availability of essential nutrients in soil.

Some saprotrophic fungi possessing the ability to decompose organic matter and release nutrients trapped within their components have evolved to develop the capacity to interact with the roots of plants, which are called mycorrhiza fungi. This interaction necessitates several physiological and morphological adaptations in both fungi and plants. The ability of sapotrophic fungi to thrive in living plants is hindered by their inability to suppress the immune systems of plants. Consequently, these fungi are confined to the decomposition of organic matter. Notably, some fungi have lost their capacity for saprotrophic growth and now rely entirely on plants for their survival.

This book discusses different aspects of fungi, such as arbuscular mycorrhizal fungi, endomycorrhizal fungi, and nanoparticles using mycorrhizal fungi for sustainable production.

This book is for researchers, students, and people who never tire of learning. I would like to thank my wife Fernanda, my daughter Maria Eduarda, and my son João Henrique for making my life happier. I also thank my deceased father Angelo Sidney Rigobelo, who lives in my mind and in my heart every day of my life.

Everlon Cid Rigobelo
Plant Production Department,
Universidade Estadual Paulista,
Campus de Jaboticabal,
Jaboticabal, São Paulo, Brazil

Section 1

Mycorrhizal Fungi

Chapter 1

Introductory Chapter: The Importance of Mycorrhiza Fungi to Sustainable Food Production

Everlon Cid Rigobelo

1. Introduction

Fungi are eukaryotic microorganisms that possess a structure similar to plant roots, known as hyphae, which are responsible for absorbing water and nutrients from the soil and providing support to the fungi. Unicellular fungi that do not produce hyphae are referred to as yeasts [1].

Fungi exhibit a vast range of metabolic and physiological capabilities, allowing them to produce an extensive array of secondary metabolites, as well as a variety of enzymes that play critical roles in the soil's biochemical cycles. Additionally, these organisms play a crucial role in increasing the availability of essential nutrients in the soil [2].

Some saprotrophic fungi, possessing the ability to decompose organic matter and release nutrients trapped within their components, have evolved to develop the capacity to interact with the roots of plants, which are called mycorrhiza fungi. This interaction necessitates several physiological and morphological adaptations in both fungi and plants. The ability of saprotrophic fungi to thrive in living plants is hindered by their inability to suppress the immune systems of plants. Consequently, these fungi were confined to the decomposition of organic matter. Notably, some fungi have lost their capacity for saprotrophic growth and now rely entirely on plants for their survival [3].

Fungi engage in a mutually beneficial partnership with the host plant, providing the latter with various advantages, while simultaneously obtaining carbohydrates and energy-rich molecules from the host [4]. Fungi are categorized into various types of mycorrhiza, based on their level of interaction with plants. One of these is Arbuscular Mycorrhizal Fungi (AMF), which establishes intricate branching structures called arbuscules within the root cells of plants, facilitating the transfer of nutrients and water from the soil to the plants through hyphae and the recycling of energy and carbohydrates from the host plant. The consequence of this interaction is increased plant growth [5].

In contrast, another type of mycorrhizal fungi is less intimate with the plant host Ectomycorrhiza Fungi (EMF). This type of mycorrhiza forms a structure known as ectosclerotium within the roots. EMF provides nutrients such as phosphorus, nitrogen, and water to plants. The relationship between plants and EMF is not totally integrated, and EMF does not colonize the entire root system [6].

Mycoheterotrophic mycorrhizae are a significant type of mycorrhizal fungi that include non-photosynthetic plant species, such as Indian or ghost pipes, and fungi.

IntechOpen

In this type of mycorrhizal association, the plant obtains nutrients solely from the fungus, which leads to a complex network of roots and fungi [7]. Another type of mycorrhizal fungus is ericoid mycorrhiza. This type of mycorrhizae is primarily associated with ericaceous plants, such as heathers and blueberries, and shares similarities with arbuscular mycorrhizae but exhibits distinct interactions with its host [1].

Orchid mycorrhizae, another type of mycorrhizae, is essential for seed germination and the early growth of many orchid species. The fungus forms dense coils around the roots of the orchid, which are crucial for its survival [8]. The existence of monotropoid mycorrhizae, a type of mycorrhizal fungus, is noteworthy. This type of mycorrhizae is unique because it is found in plants that lack chlorophyll, such as in ghost pipes. Plants rely entirely on fungi for their nutritional needs, resulting in the development of a complex root-fungus network [3].

Mycorrhizal fungi are unique in that they are influenced by different types of plants, making them a distinct type of fungus that colonizes soils worldwide. Approximately 250 species of mycorrhizal fungi are found in various families and genera. These fungi are obligate symbionts, meaning that they rely on plants to obtain energy and carbohydrate for survival. At the same time, they are essential participants in several nutrient cycles that are important for all living organisms, including nutrients such as phosphorus, nitrogen, and potassium [6].

It is widely believed that the relationship between land plants and mycorrhizal fungi has existed since the dawn of the former, which stretches back hundreds of millions of years. Among the various mutually beneficial interactions that exist between plants and microorganisms, arbuscular mycorrhizal symbiosis (AMF) is the most prevalent. Research indicates that AMF play a pivotal role in the nutrition and growth of plants, particularly under stress conditions, and help to maintain several crucial ecosystem processes. This association benefits both parties, as fungi assist plants in absorbing essential nutrients while receiving carbohydrates from them. In addition to decomposing dead organic matter, ectomycorrhizal fungi also form partnerships with forest trees and contribute to the storage of soil carbon [9].

Mycorrhizal fungi are crucial for the proper functioning of soils, element cycling, and the development of sustainable soil and land-use practices [10]. These fungi contribute significantly to sustainable agriculture by improving the soil structure, enhancing nutrient acquisition, and reducing biotic and abiotic stress, leading to increased plant growth and productivity. They form mutualistic relationships with plant roots, facilitating the efficient absorption of essential mineral nutrients, such as nitrogen, phosphorus, and potassium, and reducing dependence on chemical fertilizers. Additionally, mycorrhizal fungi can help manage plant diseases by antagonizing pathogens and can assist plants in enduring abiotic stresses, such as drought, salinity, and heavy metal toxicity. They also enhance soil quality and stability by increasing soil structure through the secretion of glomalin. The integration of mycorrhizal fungi into agricultural production offers a viable alternative for addressing the challenges faced by conventional agricultural systems. By promoting plant growth, decreasing dependence on chemical inputs, and enhancing soil health, this approach provides a sustainable solution for agriculture [11].

2. Mycorrhizal fungi and nitrogen

Nitrogen (N) is a crucial macronutrient for plants and plays a vital role in several physiological processes. The significance of nitrogen is highlighted by the fact that it is an essential component of amino acids, which are the building blocks of proteins.

Proteins are vital to the structure and function of plant cells. Nitrogen is a part of chlorophyll, the pigment responsible for photosynthesis in plants, and its deficiency can result in chlorosis and yellowing of leaves due to reduced chlorophyll synthesis. Nitrogen is also a component of nucleic acids (DNA and RNA) that provides genetic instructions for plant growth and development. Several coenzymes and adenosine triphosphate (ATP), the main cellular energy sources, also contain nitrogen [12]. It is important to note that plants require substantial amounts of nitrogen for growth; however, a significant portion of the nitrogen present in the biosphere is unavailable to them and other microorganisms. Only nitrogenase carriers, a select group of bacteria, can obtain nitrogen from the air and make it accessible to several organisms. An adequate amount of nitrogen is crucial for the vigorous vegetative growth of plants, leading to faster growth, larger leaves, and more intense green coloration. Nitrogen deficiency usually results in stunted growth, small pale leaves, and low productivity [13]. It is not possible to measure the amount of nitrogen through soil fertility analysis because of its instability. Consequently, the appropriate amount of nitrogen for crop production must be determined by considering the crop species and estimated productivity. This amount of nitrogen must be applied through mineral fertilization, and its application must be divided many times throughout crop production to reduce nitrogen loss [14].

Although mycorrhizal fungi cannot obtain atmospheric nitrogen and convert it into a form accessible to plants, they can release nitrogen that is trapped in organic matter, rendering it available to plants, thereby facilitating plant growth. Mycorrhizal fungi offer a promising alternative for sustainably supplying nitrogen to plants, as they are an effective means of addressing the increasing demand for nitrogen in agricultural production [15].

3. Mycorrhizal fungi and phosphorus

Phosphorus (P) is an indispensable element for all living organisms, as it plays a vital role in energy transfer and cell structures. Unlike nitrogen, which has a gaseous phase, phosphorus mainly exists in a solid form, making its cycle distinct. Phosphorus (P) is a crucial macronutrient for plants and is central to several physiological processes. The significance of phosphorus in plants is demonstrated by the following points: Phosphorus is a key component of adenosine triphosphate (ATP) molecules, which serve as the primary "energy currency" of cells. ATP is essential for many biochemical reactions and provides energy for various cellular functions [16].

Phosphorus is a vital component of nucleic acids, DNA and RNA, and carries the genetic information of plants. Many coenzymes and related molecules essential for many biochemical reactions contain phosphorus in their composition [17]. Phosphorus is crucial for photosynthesis and respiration processes, as well as the synthesis and breakdown of carbohydrates. Phosphorus is essential for proper root development, especially for primary roots that grow deeper into the soil. Adequate amounts of phosphorus are vital for the formation of flowers and fruits and directly affect plant production. Phosphorus aids the proper maturation of plants and seeds [18, 19].

Plants with an adequate supply of phosphorus generally exhibit enhanced resistance to diseases and various forms of abiotic stress. Conversely, phosphorus deficiency in plants may result in stunted growth and leaves that are dark in color with a bronze or purple tinge. This is attributable to the accumulation of anthocyanins and a concurrent reduction in seed and fruit production [20]. The mobility of phosphorus in the soil is limited, which means that it does not readily migrate to the roots of plants, particularly in alkaline or extremely acidic soils. Consequently, even when phosphorus is present in the soil, its availability to plants maybe diminished [21].

Arbuscular mycorrhizal fungi (AMF) are important for the uptake of phosphorus (P) by plants, as they form symbiotic relationships with plant roots, providing essential nutrients such as phosphorus and nitrogen, while obtaining carbon from the plant root system. Soils with increased levels of available phosphorus can have detrimental effects on both the diversity of microorganisms and interactions between mycorrhizae and their plant hosts. This is because high levels of available phosphorus decrease the need for plants to rely on beneficial fungi. Consequently, when there is excess available phosphorus, plants may allocate less carbon to mycorrhizae, leading to a reduction in the diversity and abundance of mycorrhizae. This decrease in AMF diversity and abundance suggested a less symbiotic relationship. On the other hand, when the soil is phosphorus deficient, the use of arbuscular mycorrhizal fungi (AMF) to provide phosphorus to the plant is a great alternative. AMF can supply phosphorus even in soils with phosphorus deficiency, thereby promoting good crop yield [22].

4. Conclusion

Mycorrhizal fungi are widespread organisms found in various regions across the globe and are closely related to a broad range of root plant species, including agricultural crops. These fungi exhibit multiple traits that promote plant growth, and numerous studies have demonstrated their ability to increase plant growth by improving the efficiency of nutrient and water uptake from the soil as well as their availability. Specifically, mycorrhizal fungi have been shown to increase the availability of nitrogen and phosphorus in plants. While these fungi cannot fix nitrogen from the atmosphere, they are capable of mineralizing organic matter, releasing nitrogen that is trapped in molecules and cellular organelles. In addition, arbuscular mycorrhizal fungi (AMF) can increase phosphorus availability in plants. However, high levels of phosphorus in the soil can decrease the interactions between the fungus and the host plant. Some studies have shown that mycorrhizal fungi can enhance plant tolerance to environmental stress. In summary, mycorrhizal fungi can be used to promote plant growth, improve nutrient utilization efficiency, reduce environmental impacts and production costs, and enhance plant tolerance to cope with stress conditions. The use of AMF is an excellent strategy to achieve sustainable production.

Author details

Everlon Cid Rigobelo
Plant Production Department, Universidade Estadual Paulista – Campus de Jaboticabal, Jaboticabal, São Paulo, Brazil

*Address all correspondence to: everlonagro@yahoo.com.br

References

[1] Parniske M. Arbuscular mycorrhiza: The mother of plant root endosymbioses. Nature Reviews Microbiology. 2008;**6**(10):763-775. DOI: 10.1038/NRMICRO1987

[2] Wang B, Qiu YL. Phylogenetic distribution and evolution of mycorrhizas in land plants. Mycorrhiza. 2006;**16**(5):299-363. DOI: 10.1007/S00572-005-0033-6

[3] Marks GC. Causal morphology and evolution of mycorrhizas. Agriculture, Ecosystems & Environment. 1991;**35**(2-3):89-104

[4] Rooney DC, Killham K, Bending GD, Baggs E, Weih M, Hodge A. Mycorrhizas and biomass crops: Opportunities for future sustainable development. Trends in Plant Science. 2009;**14**(10):542-549. DOI: 10.1016/j.tplants.2009.08.004

[5] Smith SE, Smith FA. Roles of arbuscular mycorrhizas in plant nutrition and growth: New paradigms from cellular to ecosystem scales. Annual Review of Plant Biology. 2011;**62**:227-250. DOI: 10.1146/ANNUREV-ARPLANT-042110-103846

[6] Messa VR, Savioli MR. Improving sustainable agriculture with arbuscular mycorrhizae. Rhizosphere. 2021;**19**:1-3, 100412

[7] Pozo MJ, Azcón-Aguilar C. Unraveling mycorrhiza-induced resistance. Current Opinion in Plant Biology. 2007;**10**(4):393-398. DOI: 10.1016/j.pbi.2007.05.004

[8] Sawers RJH, Gutjahr C, Paszkowski U. Cereal mycorrhiza: An ancient symbiosis in modern agriculture. Trends in Plant Science. 2008;**13**(2):93-97. DOI: 10.1016/j.tplants.2007.11.006

[9] Cavagnaro TR, Bender SF, Asghari HR, van der Heijden MGA. The role of arbuscular mycorrhizas in reducing soil nutrient loss. Trends in Plant Science. 2015;**20**(5):283-290. DOI: 10.1016/j.tplants.2015.03.004

[10] Keymer A, Gutjahr C. Cross-kingdom lipid transfer in arbuscular mycorrhiza symbiosis and beyond. Current Opinion in Plant Biology. 2018;**44**:137-144. DOI: 10.1016/j.pbi.2018.04.005

[11] Cameron DD, Neal AL, van Wees SCM, Ton J. Mycorrhiza-induced resistance: More than the sum of its parts? Trends in Plant Science. 2013;**18**(10):539-545. DOI: 10.1016/j.tplants.2013.06.004

[12] Novoa R, Loomis RS. Nitrogen and plant production. Plant and Soil. 1981;**58**(1-3):177-204. DOI: 10.1007/BF02180053/METRICS

[13] Sheoran S, Kumar S, Kumar P, Meena RS, Rakshit S. Nitrogen fixation in maize: Breeding opportunities. Theoretical and Applied Genetics. 2021;**134**(5):1263-1280. DOI: 10.1007/S00122-021-03791-5

[14] Hedin LO, Brookshire ENJ, Menge DNL, Barron AR. The nitrogen paradox in tropical forest ecosystems. Annual Review of Ecology, Evolution, and Systematics. 2009;**40**:613-635. DOI: 10.1146/ANNUREV.ECOLSYS.37.091305.110246

[15] Jansa J, Forczek ST, Rozmoš M, Püschel D, Bukovská P, Hršelová H. Arbuscular mycorrhiza and soil organic nitrogen: Network of players and interactions. Chemical and

Biological Technologies in Agriculture. 2019;**6**(1):1-10

[16] Adnan M, Fahad S, Zamin M, Shah S, Mian IA, Danish S, et al. Coupling phosphate-solubilizing bacteria with phosphorus supplements improve maize phosphorus acquisition and growth under lime induced salinity stress. Plants. 2020;**9**(7):1-18. DOI: 10.3390/plants9070900

[17] Bindraban PS, Dimkpa CO, Pandey R. Exploring phosphorus fertilizers and fertilization strategies for improved human and environmental health. Biology and Fertility of Soils. 2020;**56**(3):299-317. DOI: 10.1007/S00374-019-01430-2/FIGURES/4

[18] Malhotra H, Vandana, Sharma S, Pandey R. Phosphorus nutrition: Plant growth in response to deficiency and excess. Plant Nutrients and Abiotic Stress Tolerance. 2018;**1**:171-190

[19] Malhotra M, Srivastava S. Stress-responsive indole-3-acetic acid biosynthesis by Azospirillum brasilense SM and its ability to modulate plant growth. European Journal of Soil Biology. 2009;**45**(1):73-80. DOI: 10.1016/j.ejsobi.2008.05.006

[20] Parvin S, Van Geel M, Yeasmin T, Verbruggen E, Honnay O. Effects of single and multiple species inocula of arbuscular mycorrhizal fungi on the salinity tolerance of a Bangladeshi rice (Oryza sativa L.) cultivar. Mycorrhiza. 2020;**30**(4):431-444. DOI: 10.1007/s00572-020-00957-9

[21] Pereira NCM, Galindo FS, Gazola RPD, Dupas E, Rosa PAL, Mortinho ES, et al. Corn yield and phosphorus use efficiency response to phosphorus rates associated with plant growth promoting bacteria. Frontiers in Environmental Science. 2020;**8**(40):1-12

[22] Ma J, Xie Y, Sun J, Zou P, Ma S, Yuan Y, et al. Co-application of chitooligosaccharides and arbuscular mycorrhiza fungi reduced greenhouse gas fluxes in saline soil by improving the rhizosphere microecology of soybean. Journal of Environmental Management. 2023;**345**:118836. DOI: 10.1016/j.jenvman.2023.118836

Chapter 2

Outlooks of Nanotechnology with Mycorrhizae

Ban Taha Mohammed

Abstract

Mycorrhizae play a vital role in providing plants with essential macro and micro-mineral elements and protecting them from pathogen infections. Enhancing the plant's resistance to environmental stress like drought, salinity, and heavy metals, along with enhancing soil structure through the secretion of glycoprotein compounds known as Glumalin, are some benefits. Additionally, plants treated with mineral nanoparticles and mycorrhiza exhibit improved growth, yield, and biochemical characteristics. Also, the plants treated with mineral nanoparticles and mycorrhiza showed better growth, yield, and biochemical properties. Mycorrhiza can also be used as a base material for the synthesis of nanoparticles under green synthesis mode. Thus, nanotechnology and biofertilization are steps friendly environmental.

Keywords: nanofertilizers, mycorrhizae, micro-mineral elements, bio fertilization, green-nanotechnology

1. Introduction

Mycorrhizae are fungi that form a symbiotic benign relationship with higher plants [1]. These associations vary widely in functions and structure, and there are several types of mycorrhizae, but the most abundant is arbuscular mycorrhizal (AM) [2]. It can supply nutrients to plants, even when levels are low in the soil, particularly vital nutrients like phosphorus (P) and nitrogen (N), as well as other elements like copper, zinc, and iron. In return, plants allocate 10–20% of their photosynthesis to fungi—mycorrhizae shield roots from harmful fungi in the soil. Arbuscular mycorrhizal fungi can establish a symbiotic bond with 85% of plant families, contributing significantly to ecosystem sustainability [3]. AM fungi are obligate mutualists belonging to the phylum Glomeromycota [4]. Fungal hyphae enter root epidermal cells and proceed to cortex cells, where they form vesicles, arbuscles, and coils, establishing a network of hyphae outside the root (extra-radical hyphae) that enhances soil texture and water relations within the plant's root system [5]. AM fungi produce a type of glycoprotein named Glyoxalin, which contributes to alleviating soil degradation and inoculating plants. AM fungi give them the ability to withstand extreme conditions of salinity, drought, temperature, and lack of mineral nutrients [6]. With these good qualities, mycorrhizae have entered organic agriculture, but they are insufficient and unrewarding in all regions and under all circumstances, and other supplements must be used, such as mineral and organic fertilizers, in addition to all types of control against diseases and insects. Nanotechnology has entered every detail of life, including agriculture, to benefit from its potential to solve many problems of environmental pollution, lack of nutritional elements and minerals, and increasing production and yields. Both humans and plants need nutritional elements for growth and development, and these elements are absorbed through the root plants and transported to all parts of the plant, so that humans can eat them as suitable food [1].

In pea plants, the synergistic activity between nanoparticles and mycorrhizae caused rising of acid phosphatase, alkaline phosphatase, catalase, and peroxidase as biochemical parameters. The combination that is used is Nano-$ZnFe_2O_4$ and arbuscular mycorrhiza (AM), in addition to improving N, P, K, and Mg uptakes [2]. The coupling between $ZnFe_2O_4$ NPs and AM fungi in nano agricultural applications Improved the physiological, growth, and biochemical parameters of pea plants [3]. The use of nanofoliar fertilizers and nanofertilizers extracted from marine organisms with mycorrhizal fungi enhanced the nutrient concentration, growth, and production yield of wheat [4]. The combined application of foliar spray AM fungi (250 g inoculum/vine) and seaweed extract (1%) with nano-zinc oxide particles (50 ppm) three times after the bud burst stage on sweet grapevines resulted in improved vegetable growth traits, nutritional status, yield, and cluster quality attributes. The interaction between mycorrhizae, nanocomposites, and their challenges, particularly in combating various microorganisms like fungi, will be the central focus of this chapter, highlighting both the challenges and advantages.

2. Management of drought stress using nanofertilizers and mycorrhizae

In arid and semi-arid regions, successive droughts resulting from climate change have led to abiotic stress, causing poor plant growth and low productivity [7]. Lack of soil moisture can have limitation effects on hydraulic and nutritional conductivity, especially phosphorus (P), potassium (K), nitrogen (N), ionic balance, and gas exchange; it damages photosynthetic pigments and reduces the photochemical efficiency of photosystem II (PSII) [8–10]. In addition, reactive oxygen species (ROS) are produced and can damage cell structures such as carbohydrates, nucleic acids, lipids, and proteins and alter their functions, creating oxidative stress [11]. To maintain a plant's water balance, stomata arrange leaf diffusive conductance to control the flow of gases between the stomatal aperture leaf and the atmosphere, reducing the diffusion of carbon dioxide (CO_2), leading to a decline in photosynthetic performances and chlorophyll pigment concentration, as well as accumulation of primary metabolites such as soluble sugars and proteins [12]. For these reasons, the combination of plants and AM fungi could be an appropriate strategy for the sustainable management of aridity stress. In pot experiments, wheat plants (*Triticum aestivum*) inoculated with arbuscular mycorrhizal fungi (AMF) showed an increase in relative water content in both leaves and soil, which indicates resistance to drought stress through the ability of AMF to penetrate deep into the soil to provide the necessary moisture. The AMF plant colonization was less severe than the drought stress damage to the photochemistry, function, and structure of PSII and PSI [13]. Nanofertilizers are remarkable tools in agriculture that help to progress crop growth, yield, and quality parameters [14]. The comparison between the nano and conventional mineral nutrients fertilizers results in increased nutrient cause toxicity and inhibits crop growth, compared to the use of fertilizers and nano-nutrients, which improve the growth and yield without toxicity [15]. Nanofertilizers supply and expand the surface area to perform many types of metabolic reactions, leading to an increased rate of photosynthesis, dry matter, and yield of the crop [14]. The optimum dose and concentration applied are important to get the maximum growth and yield of the crop; otherwise, they also have an inhibitory effect if they exceed the optimum one [15]. The combined treatment carbon nanoparticles, arbuscular mycorrhiza, and compost (Com-AMF-CNPs) on maize plant crops under drought stress increased the levels of oxalic and succinic acids (by 100%) as indicators of enhanced amino acids level such as the antioxidant and omso-protectant proline, increases in fatty acids such as myristic, palmitic, arachidic, docosanoic, and pentacosanoic (110, 30, 100, and 130%, respectively), compared with untreated plants, polyamine biosynthesis amino acids and increased spermine and spermidine synthases, ornithine decarboxylase, and adenosyl methionine decarboxylase in treated maize [16]. Drought stress significantly reduced maize growth and photosynthesis and altered metabolism [17]. Arbuscular mycorrhizal fungi (AMF) and titanium dioxide (TiO_2) have been proven to mitigate the impacts of drought stress on plants (*Salvia officinalis* L.). Combining biofertilizer and nanoparticles (TiO_2 + AMF) offers a sustainable approach to enhancing essential oils (EO), boosting productivity, and improving the quality of sage plants under drought-stress conditions [18]. Transmission electron microscopy and X-ray analyses were used to detect the combination between Ferrous nanoparticles (FeNPs) and AMF (Glomus) to enhance photosynthesis and water efficiency, in addition to increasing uptake of 'Fe-NPs,' which means the presence of elements in seeds and roots enhanced the growth-promoting and drought-tolerant [19].

3. Reducing salinity stress using nanomaterials and mycorrhizae

Salt stress is a pioneer agricultural trouble all over the world; reducing salinity stress using nanomaterials and mycorrhizae has an effect on the functioning of growth and physiology of crop plants [20]. Soil salinity affects the rate of plant growth by reducing the rate of photosynthesis and stomatal conductivity and inhibiting antioxidant enzymes [21]. Under salinity stress conditions, using arbuscular mycorrhizal fungi (AMF) enhances crop plant growth and physiological performance, increasing plant biomass, photosynthetic activity, water potential, and selective uptake of nutrients [22]. As a response to salinity stress, AMF increases antioxidant defense mechanisms and promotes salinity tolerance in crop plants [23]. AMF plays an important role in plant cell osmotic balance by regulating the Na and K ratio and enhancing the synthesis of many osmolytes such as proline and glycine betaine [24]. AM fungi may be able to keep plant cells from salt stress, but there is a decline in propagule production and AM colonization under high salinity conditions [25]. Therefore, the combination of nanoparticles and AMF resolved the problem, and the AMF community co-occurrence network complexity was enhanced by silver nanoparticle AgNPs [26]. To improve production under salinity conditions, it is recommended to use nano-zinc oxide and mycorrhizae [27]. In the salinity condition, the interaction between nano-zinc oxide and AM fungi showed significantly increased K^+, reduced Na^+ uptake, and increased wheat yield [28]. In addition, watering the wheat plant with different levels of salt water may have an effect on the physiological properties and yield, but the use of the interaction between mineral and nano-zinc and mycorrhizal fungi improved the physiological properties, yield, concentration of some nutrients, the K/Na ratio, and the protein percentage in the crop grains [29]. The combination of nano-zinc fertilizer and mycorrhizae has a notable impact on the activity of (CAT), (SOD), and (POD) enzymes when exposed to varying levels of saline water during irrigation. This impact becomes more pronounced as the salinity levels in the irrigation water rise (**Tables 1–3**) [30]. In a greenhouse environment, silicon nanoparticles (Si-NPs) at different concentrations (0, 30, and 60 mg/L) along with various biofertilizers were examined for their potential to enhance wheat's resistance to salinity stress by evaluating antioxidant activity. The study involved different salt concentrations (0, 35, 70, and 105 mM), and the outcomes indicated that bio-fertilizers and Si-NPs boosted parameters, including plant leaf water potential, soluble protein, soluble sugar, catalase, peroxidase, and polyphenol oxidase levels. The application of both single and combined bio-fertilizers with Si-NPs showed improvements in physiological aspects such as stomatal conductance,

Mycorrhiza M (20 g/pot)		**Type of zinc Zn (15 mg L^{-1})**	**Salinity levels of irrigation water S (dS m^{-1})**				**Average of mycorrhiza**
			S1	**S2**	**S3**	**S4**	
M0		Zn0	4.97	4.80	6.11	10.19	6.19
		Zn1	3.70	4.23	4.78	7.86	
		Zn2	3.89	6.88	8.52	8.36	
M1		Zn0	5.03	5.35	8.60	8.80	7.03
		Zn1	3.67	6.48	9.16	8.93	
		Zn2	4.89	6.62	7.39	9.41	
Salinity levels of irrigation rate			4.36	5.73	7.43	8.92	
Zinc rate			Zn0	Zn1	Zn2		
			6.73	6.10	7.00		
M. Zn			Zn0	Zn1	Zn2		
M0			6.52	5.14	6.91		
M1			6.95	7.06	7.08		
M. S			S1	S2	S3	S4	
M0			4.19	5.30	6.47	8.80	
M1			4.53	6.15	8.38	9.05	
S. Zn			S1	S2	S3	S4	
Zn0			5.00	5.07	7.35	9.49	
Zn1			3.68	5.36	6.97	8.39	
Zn2			4.39	6.75	7.96	8.89	
L.S.D0.05	M	Zn	S	Zn. M	S.M	S. Zn	S. Zn.M
	0.405	0.496	0.573	0.702	0.810	0.992	1.403

M0 = without mycorrhiza *
M1 = with mycorrhiza
Zn0 = without zinc
Zn1 = mineral zinc
Zn2 = nano-zinc

S1 = 2 dS m^{-1}
S2= 4 dS m^{-1}
S3 = 6 dS m^{-1}
S4 = 8 dS m^{-1}

Table 1.
The interaction between mycorrhiza, zinc, and different levels of saline water in CAT enzyme (unit g. protein^{-1} fresh weight) of wheat crop.

Mycorrhiza M (20 g/pot)	**Type of zinc Zn (15 mg L^{-1})**		**Salinity levels of irrigation water S (dS m^{-1})**				**Average of mycorrhiza**
			S1	**S2**	**S3**	**S4**	
M0	Zn0		3.81	3.96	3.15	5.21	3.75
	Zn1		2.91	3.57	3.99	4.36	
	Zn2		3.08	3.20	4.75	3.03	
M1	Zn0		4.67	3.31	3.19	3.46	3.28
	Zn1		1.20	2.91	3.01	3.76	
	Zn2		3.78	3.43	3.00	3.69	
Salinity levels of irrigation rate			3.24	3.40	3.52	3.92	
Zinc rate			Zn0	Zn1	Zn2		
			3.84	3.21	3.49		
M. Zn			Zn0	Zn1	Zn2		
M0			3.27	3.58	3.96		
M1			3.21	3.21	3.05		
M. S			S1	S2	S3	S4	
M0			3.27	3.58	3.96	4.20	
M1			3.21	3.21	3.05	3.64	
S. Zn			S1	S2	S3	S4	
Zn0			4.24	3.64	3.17	4.33	
Zn1			2.05	3.24	3.50	4.06	
Zn2			3.43	3.31	3.88	3.36	
L.S.D0.05	M	Zn	S	Zn. M	S.M	S. Zn	S. Zn.M
	0.420	0.514	0.594	0.727	0.840	1.029	1.455

M0 = without mycorrhiza *
M1 = with mycorrhiza
Zn0 = without zinc
Zn1 = mineral zinc
Zn2 = nano-zinc

S1 = 2 dS m^{-1}
S2 = 4 dS m^{-1}
S3 = 6 dS m^{-1}
S4 = 8 dS m^{-1}

Table 2.
The interaction between mycorrhiza, zinc, and different levels of saline water in SOD enzyme (unit g. protein^{-1} fresh weight) of wheat crop.

electrical conductivity, electrolytic leakage, and proline content. Furthermore, the use of Si-NPs and bio-fertilizers led to increased wheat grain yield under salinity stress conditions by enhancing physiological and biochemical characteristics. The most favorable results in terms of grain yield and reduced malondialdehyde and hydrogen peroxide levels were observed with the combined application of biofertilizers and Si-NPs at a concentration of 60 mg/L under normal salinity levels [31]. Salinity stress limits crop production. The use of silicon dioxide nanoparticles (SiO_2-NPs) along with the inoculation of arbuscular mycorrhizal fungi (AMF) and plant growth-promoting rhizobacteria (PGPR) with wheat helps mitigate the negative impacts of salinity stress. This combined approach enhances physiological parameters

Mycorrhiza M (20 g/pot)	Type of zinc Zn (15 mg L^{-1})	Salinity levels of irrigation water S (dS m^{-1})				Average of mycorrhiza
		S1	S2	S3	S4	
M0	Zn0	1.979	2.435	2.328	2.917	2.304
	Zn1	1.837	2.421	2.523	2.373	
	Zn2	1.876	2.273	2.238	2.453	
M1	Zn0	2.492	2.452	2.135	2.381	2.175
	Zn1	1.463	2.129	2.574	2.964	
	Zn2	1.441	1.723	1.902	2.450	
Salinity levels of irrigation rate		1.848	2.239	2.283	2.590	
Zinc rate		Zn0	Zn1	Zn2		
		2.390	2.285	2.045		
M. Zn		Zn0	Zn1	Zn2		
M0		2.415	2.288	2.210		
M1		2.365	2.282	1.879		
M. S		S1	S2	S3	S4	
M0		1.897	2.376	2.363	2.581	
M1		1.799	2.101	2.203	2.598	
S. Zn		S1	S2	S3	S4	
Zn0		2.235	2.444	2.231	2.649	
Zn1		1.650	2.275	2.548	2.668	
Zn2		1.658	1.998	2.070	2.452	

L.S.D0.05	M	Zn	S	Zn. M	S.M	S. Zn	S. Zn.M
	n.s	0.2325	0.2684	0.3288	0.3796	0.4649	0.6575

*M0 = without mycorrhiza
M1 = with mycorrhiza
Zn0 = without zinc
Zn1 = mineral zinc
Zn2 = nano-zinc

S1 = 2 dS m^{-1}
S2 = 4 dS m^{-1}
S3 = 6 dS m^{-1}
S4 = 8 dS m^{-1}

Table 3.
The interaction between mycorrhiza, zinc, and different levels of saline water in POD enzyme (unit g. protein^{-1} fresh weight) of wheat crop.

such as leaf area index, chlorophyll fluorescence, anthocyanin, and chlorophyll index. Coinoculation of AMF and PGPR, along with 60 mg/L SiO_2-NPs, leads to improved Si, P, and K+ uptake, reduced Na+ uptake, and ultimately, increased grain yield [32].

4. Bioremediation of heavy metals using nanomaterials and mycorrhizae

Heavy metal elements like copper (Cu), lead (Pb), mercury (Hg), and zinc (Zn) have made their way into the human living environment through various pathways such as atmospheric precipitation, dust, transportation, and activities like agriculture, industrial wastewater, and vehicle emissions [33–35], resulting in serious heavy metal pollution that can accumulate in crops and animals, affecting their growth and poisoning and eventually entering the human body through the food chain and food web, endangering human health and even life [36]. A variety of diseases can arise from the accumulation of heavy metal elements in the human body, such as cancers [37]. Improving soils and waste contaminated with heavy metals is crucial for achieving a clean and sustainable environment. Arbuscular mycorrhizal fungi (AMF) play a role in alleviating the harmful effects of different pollutants on plants [38]. Copper oxide nanoparticles (nano-CuO) are considered a growing environmental concern for *Canna indica* seedlings. In this study, AMF inoculation effectively alleviates the stress caused by nano-CuO by boosting antioxidant enzyme activities and reducing ROS levels in both the leaves and roots of *C. indica*. This led to an upregulation of antioxidant genes, a decrease in Cu levels within the seedlings, and an increase in the expression of Cu transport genes and metallothionein genes [39]. Engineered nanomaterials (ENMs) are on the rise across various sectors, leading to a significant risk of ENMs entering the natural environment. Rhizomicroorganisms, such as arbuscular mycorrhizal fungi (AMF), play a crucial role in immobilizing ENMs, reducing phytoaccumulation, enhancing host plant growth, triggering system-wide protection against ENMs stress, and indirectly mitigating the harmful effects of ENMs on host plants. The combined action of plants and AMF could offer a cost-effective and environmentally friendly approach to controlling potential ENM contamination sustainably [40]. AMF's ability to reduce the negative effects of elevated AgNP levels, such as increasing plant growth and improving ecological behaviors, was enhanced, resulting in lower Ag levels and reduced antioxidant enzyme activity in plants [41]. Maize (*Zea mays* L.) exhibited varying responses when exposed to different concentrations of Silver nanoparticles (AgNPs) (0, 0.025, 0.25, and 2.5 mg/kg) both in the presence and absence of arbuscular mycorrhizal (AM) fungi and soil microorganisms. The incorporation of AM fungi was found to play a crucial role in ameliorating the adverse effects induced by AgNPs phytotoxicity, thereby enhancing the mycorrhizal colonization, biomass, and phosphorus (P) uptake in maize plants. Furthermore, the inoculation of AM fungi was associated with a reduction in the toxicity of AgNPs toward soil microbial communities, leading to an increase in bacterial diversity and prompting alterations in the composition of the soil bacterial community. This shift in the soil bacterial community structure was attributed to the modulation of potential Ag transporter gene expressions and the collaborative interactions between the AM fungi and soil bacterial community [42]. Inoculating Spearmint plants with arbuscular mycorrhizal (AM) helped alleviate Cu toxicity caused by spraying 0.66 and 1.05 mg of elemental Cu per pot from $Cu(OH)_2$ nanowires, bulk $Cu(OH)_2$ (Kocide 3000), or ionic $CuSO_4$. This study illustrates that interactions between plants and microbes could enhance crop productivity and offer environmental advantages, highlighting the importance of assessing risks in crop systems that involve symbiotic microorganisms [43]. Soil microorganisms play a vital role in explaining the impact of iron oxide nanoparticles (Fe_3O_4NPs) across different concentrations (0.1, 1.0, and 10.0 mg/kg) on maize plants with or without AMF inoculation. No significant changes in soil biota or dissolved organic C (DOC) were observed in AMF-inoculated treatments. However, variations were noted in content levels under Fe_3O_4NPs treatment alone. These findings suggest that AMF can modify the effects of Fe_3O_4NPs on soil microorganisms, potentially through influencing plant growth and the release of organic matter from plant roots, as evidenced by the impact on DOC levels due to AMF [44].

5. Negatively influence nanomaterials interaction of mycorrhizae

Nanoparticles interact with mycorrhizae in various ways, some promoting colonization by AMF, while others hinder it [45]. Therefore, more detailed information is required on how arbuscular mycorrhizal fungi (AMF) interact with nanoparticles. High levels of Fe_3O_4NPs can disrupt the symbiotic relationship between AM fungi and plants, leading to carbon and phosphorus cycling deterioration in the soil. These impacts are detrimental to crop yield and soil fertility [46]. The ecological system influences the growth of mycorrhizal clover, with 0.01 mg/kg of AgNPs inhibiting growth, while 0.1 mg/kg of AgNPs leads to increased AgNP content. This affects the ability of AMF to alleviate AgNP stress by reducing Ag content and the activity of antioxidant enzymes in plants [41]. The ecosystem significantly influences the growth of mycorrhiza. A concentration of 0.01 mg/kg of nanosilver AgNPs suppresses mycorrhizal growth, while higher concentrations such as 0.1 mg/kg have adverse effects on the AM fungal colonization in the roots of Corn plants (*Zea mays* L.). This leads to reduced nutrient absorption, impacting the phosphorus cycle and uptake in the root zone, ultimately hindering plant growth and soil fertility [47].

6. Conclusions

Arbuscular mycorrhizal fungi (AMF) always associate with the roots of higher plants and form a mutualistic symbiosis with the roots of over 90% of plant species. The results of the study showed that the employment of nanofertilizers and mycorrhizal fungi had significant improvements in some physiological and chemical characteristics of plants and decreased the negative effects of drought stress, as well as their positive effect on antioxidants and cell death under salinity stress. Recently, a revolution has been created in the world of nanotechnology, and its interaction with mycorrhizae has resulted in beneficial plant growth, biological enrichment, and improved connection of mycorrhizae with plants. Other challenges were negative, given that increasing the concentration of nanomaterials may lead to inhibiting mycorrhizal growth and infecting plant roots, which is worth studying in order to find the best that deserves study in order to find the best ways for this relationship. Hence, in this chapter, we concentrate on the effect of various nanoparticles combined with AMF on some physiological characteristics of plants under stress and the role of nanoparticles in positive and negative relationships.

Acknowledgements

To the administration of the University of Kerbala, and the College of Education for Pure Sciences for the support.

Conflict of interest

The authors declare no conflict of interest.

Dedication

To my dear ones.
My late husband, Professor Dr. Abdoun Hashim Alwan, Thanks and mercy.
My dear children.
Zainab, Zaid and Zeki.

Author details

Ban Taha Mohammed
University of Kerbala, Karbala, Iraq

*Address all correspondence to: bantmh@gmail.com; ban.taha@uokerbala.edu.iq

References

[1] Khaliq A, Perveen S, Alamer KH, Zia Ul Haq M, Rafique Z, Alsudays IM, et al. Arbuscular mycorrhizal fungi symbiosis to enhance plant–soil interaction. Sustainability. 2022;**14**(13):7840

[2] Brundrett M. Diversity and classification of mycorrhizal associations. Biological Reviews. 2004;**79**(3):473-495

[3] Montesinos-Navarro A, Segarra-Moragues JG, Valiente-Banuet A, Verdú M. Plant facilitation occurs between species differing in their associated arbuscular mycorrhizal fungi. The New Phytologist. 2012;**196**(3):835-844

[4] Willis A, Rodrigues BF, Harris PJC. The ecology of arbuscular mycorrhizal fungi. CRC Critical Review in Plant Sciences. 2013;**32**(1):1-20

[5] Wang F, Zhang L, Zhou J, Rengel Z, George TS, Feng G. Exploring the secrets of hyphosphere of arbuscular mycorrhizal fungi: Processes and ecological functions. Plant and Soil. 2022;**481**(1):1-22

[6] Prasad K, Khare A, Rawat P. Glomic role in global sustainlomalin arbuscular mycorrhizal fungal reproduction, lifestyle and dynaable agriculture for future generation. In: Handbook of Fungal Reprod Growth. 1st ed. London, United Kingdom; 2022. pp. 1-22. DOI: 10.05772/intechopen95215.ch6

[7] Boutasknit A, Ait-el-mokhtar M, Fassih B, Ben-laouane R. Effect of arbuscular mycorrhizal fungi and rock phosphate on growth, physiology, and biochemistry of carob under water stress and after rehydration in vermicompost-amended soil. Metabolites. 2024;**14**(4):202. DOI: 10.3390/metabo14040202

[8] Reed C, Wang R, Zhang Z, Wang H, Chen Y. Soil water deficit reduced root hydraulic conductivity of common reed (Phragmites australis). Plants. 2023;**12**(20):3543. DOI: 10.3390/plants12203543

[9] Khan F, Siddique AB, Shabala S, Zhou M. Phosphorus plays key roles in regulating plants' physiological responses to abiotic stresses. Plants. 2023;**12**(15):2861. DOI: 10.3390/plants12152861

[10] Lu C, Zhang J. Effects of water stress on photosystem II photochemistry and its thermostability in wheat plants. Journal of Experimental Botany. 1999;**50**(336):1199-1206

[11] Birben E, Sahiner UM, Sackesen C, Erzurum S, Kalayci O. Oxidative stress and antioxidant defense. World Allergy Organization Journal. 2012;**5**:9-19

[12] Buckley TN. The control of stomata by water balance. The New Phytologist. 2005;**168**(2):275-292

[13] Mathur S, Tomar RS, Jajoo A. Arbuscular mycorrhizal fungi (AMF) protects photosynthetic apparatus of wheat under drought stress. Photosynthesis Research. 2019;**139**:227-238

[14] Singh MD. Nano-fertilizers is a new way to increase nutrients use efficiency in crop production. International Journal of Agricultural Science. 2017;**9**(7):975-3710

[15] Zhang Y, Goss GG. Nanotechnology in agriculture: Comparison of the toxicity between conventional and nano-based agrochemicals on non-target aquatic species. Journal of Hazardous Materials. 2022;**439**:129559

[16] Alsherif EA, Almaghrabi O, Elazzazy AM, Abdel-Mawgoud M, Beemster GTS, AbdElgawad H. Carbon nanoparticles improve the effect of compost and arbuscular mycorrhizal fungi in drought-stressed corn cultivation. Plant Physiology and Biochemistry. 2023;**194**:29-40

[17] Alsherif EA, Almaghrabi O, Elazzazy AM, Abdel-Mawgoud M, Beemster GTS, Sobrinho RL, et al. How carbon nanoparticles, arbuscular mycorrhiza, and compost mitigate drought stress in maize plant: A growth and biochemical study. Plants. 2022;**11**(23):3324

[18] Ostadi A, Javanmard A, Amani Machiani M, Sadeghpour A, Maggi F, Nouraein M, et al. Co-application of TiO_2 nanoparticles and arbuscular mycorrhizal fungi improves essential oil quantity and quality of sage (*Salvia officinalis* L.) in drought stress conditions. Plants. 2022;**11**(13):1659

[19] Naseer M, Zhu Y, Li FM, Yang YM, Wang S, Xiong YC. Nano-enabled improvements of growth and colonization rate in wheat inoculated with arbuscular mycorrhizal fungi. Environmental Pollution. 2022;**295**:118724

[20] Majeed A, Muhammad Z. Salinity: A major agricultural problem—Causes, impacts on crop productivity and management strategies. In: Hasanuzzaman M, Hakeem K, Nahar K, Alharby H, editors. Handbook of Plant Abiotic Stress Tolerance. Cham: Springer; 2019. pp. 83-99. DOI: 10.1007/978-3-030-06118-0_3

[21] Safdar H, Amin A, Shafiq Y, Ali A, Yasin R, Shoukat A, et al. A review: Impact of salinity on plant growth. Natural Science. 2019;**17**(1):34-40

[22] Beltrano J, Ruscitti M, Arango MC, Ronco M. Effects of arbuscular mycorrhiza inoculation on plant growth, biological and physiological parameters and mineral nutrition in pepper grown under different salinity and p levels. Journal of Soil Science and Plant Nutrition. 2013;**13**(1):123-141

[23] Zhan F, Li B, Jiang M, Yue X, He Y, Xia Y, et al. Arbuscular mycorrhizal fungi enhance antioxidant defense in the leaves and the retention of heavy metals in the roots of maize. Environmental Science and Pollution Research. 2018;**25**:24338-24347

[24] Wang J, Yuan J, Ren Q, Zhang B, Zhang J, Huang R, et al. Arbuscular mycorrhizal fungi enhanced salt tolerance of Gleditsia sinensis by modulating antioxidant activity, ion balance and P/N ratio. Plant Growth Regulation. 2022;**97**(1):33-49

[25] Borde M, Dudhane M, Kulkarni M. Role of arbuscular mycorrhizal fungi (AMF) in salinity tolerance and growth response in plants under salt stress conditions. In: Varma A, Prasad R, Tuteja N, editors. Handbook of Mycorrhiza-Eco-Physiology, Secondary Metabolites Nanomaterials. Cham: Springer; 2017. pp. 71-86. DOI: 10.1007/978-3-319-57849-1_5

[26] Chen L, Xiao Y. Silver nanoparticles and arbuscular mycorrhizal fungi influence Trifolium repen root-associated AMF community structure and its co-occurrence pattern. Scientia Horticulturae (Amsterdam). 2023;**320**:112232

[27] Online IP. Journal of crop nutrition. Science. 2017;**3**:3

[28] Seyed Sharifi R, Khalilzadeh R, Soltanmoradi S. Effects of mycorrhizal fungi and nano zinc oxide on seed

yield, Na+ and K+ content of wheat (*Triticum aestivum* L.) under salinity stress. Journal of Crop Nutrition Science. 2017;**3**(4):40-53

[29] Aljenaby HK, Mohammed BT, Al-Semmak QH. The interaction between the mineral and nano zinc with mycorrhiza on the concentration of some nutrients, the ratio of K/Na and the proportion of protein in grains of wheat (*Triticum aestivum* L.) irrigated with saline water. NeuroQuantology. 2022;**20**(12):1183

[30] Aljenaby HK, Al-Semmak QH, Mohammed BT. The interaction between nano zinc and mycorrhiza on some wheat enzymes activity under saline conditions. Jundishapur Journal of Microbiology. 2022;**15**(2):539-557

[31] Ahmadi-Nouraldinvand F, Sharifi RS, Siadat SA, Khalilzadeh R. Reduction of salinity stress in wheat through seed bio-priming with mycorrhiza and growth-promoting bacteria and its effect on physiological traits and plant antioxidant activity with silicon nanoparticles application. Silicon. 2023;**15**(16):6813-6824

[32] Ahmadi-Nouraldinvand F, Sharifi RS, Siadat SA, Khalilzadeh R. Fascinating role of silicon dioxide nanoparticles and co-inoculation of mycorrhiza and Rhizobacteria to combat NaCl stress: Changes in physiological characteristics, uptake of nutrient elements, and enhancing photosystem II activities in wheat. Journal of Soil Science and Plant Nutrition. 2024;**24**:277-294

[33] Khalef RN, Hassan AI, Saleh HM. Heavy metal's environmental impact. In: Environmental Impact and Remediation of Heavy Metals. London, UK: IntechOpen; 2022

[34] Cimboláková I, Uher I, Laktičová KV, Vargová M, Kimáková T, Papajová I. Heavy metals and the environment. Environmental factors Affect Human Health. 2019;**1**:29

[35] Rahman Z, Singh VP. The relative impact of toxic heavy metals (THMs) (arsenic (As), cadmium (Cd), chromium (Cr)(VI), mercury (Hg), and lead (Pb)) on the total environment: An overview. Environmental Monitoring and Assessment. 2019;**191**:1-21

[36] Mohammed AS, Kapri A, Goel R. Heavy metal pollution: Source, impact, and remedies. In: Khan M, Zaidi A, Goel R, Musarrat J, editors. Handbook of Biomanagement of Metal-Contaminated Soils. Environmental Pollution. Vol. 20. Dordrecht: Springer; 2011. pp. 1-28. DOI: 10.1007/978-94-007-1914-9_1Met soils

[37] Budi HS, Catalan Opulencia MJ, Afra A, Abdelbasset WK, Abdullaev D, Majdi A, et al. Source, toxicity and carcinogenic health risk assessment of heavy metals. Reviews on Environmental Health. 2024;**39**(1):77-90

[38] Herath B, Madushan KWA, Lakmali JPD, Yapa PN. Arbuscular mycorrhizal fungi as a potential tool for bioremediation of heavy metals in contaminated soil. World Journal of Advanced Research and Reviews. 2021;**10**(3):217-228

[39] Luo J, Yan Q, Yang G, Wang Y. Impact of the arbuscular mycorrhizal fungus Funneliformis mosseae on the physiological and defence responses of Canna indica to copper oxide nanoparticles stress. Journal of Fungi. 2022;**8**(5):513

[40] Wang L, Yang D, Ma F, Wang G, You Y. Recent advances in responses of arbuscular mycorrhizal fungi-plant symbiosis to engineered nanoparticles. Chemosphere. 2022;**286**:131644

[41] Feng Y, Cui X, He S, Dong G, Chen M, Wang J, et al. The role of metal nanoparticles in influencing arbuscular mycorrhizal fungi effects on plant growth. Environmental Science & Technology. 2013;**47**(16):9496-9504

[42] Cao J, Feng Y, Lin X, Wang J. A beneficial role of arbuscular mycorrhizal fungi in influencing the effects of silver nanoparticles on plant-microbe systems in a soil matrix. Environmental Science and Pollution Research. 2020;**27**(11):11782-11796

[43] Apodaca SA, Cota-Ruiz K, Hernandez-Viezcas JA, Gardea-Torresdey JL. Arbuscular mycorrhizal fungi alleviate phytotoxic effects of copper-based nanoparticles/compounds in spearmint (*Mentha spicata*). ACS Agricultural Science & Technology. 2022;**2**(3):661-670

[44] Cao J, Feng Y, Lin X, Wang J. Arbuscular mycorrhizal fungi alleviate the negative effects of iron oxide nanoparticles on bacterial community in rhizospheric soils. Frontiers in Environmental Science. 2016;**4**:10

[45] Ingle A, Rathod D, Varma A, Rai M. Understanding the mycorrhiza-nanoparticles interaction. In: Varma A, Prasad R, Tuteja N, editors. Handbook of Mycorrhiza-Eco-Physiology, Secondary Metabolites, Nanomaterials. Cham: Springer; 2017. pp. 311-324. DOI: 10.1007/978-3-319-57849-1_18

[46] Cao J, Feng Y, Lin X, Wang J, Xie X. Iron oxide magnetic nanoparticles deteriorate the mutual interaction between arbuscular mycorrhizal fungi and plant. Journal of Soils and Sediments. 2017;**17**:841-851

[47] Cao J, Feng Y, He S, Lin X. Silver nanoparticles deteriorate the mutual interaction between maize (*Zea mays* L.) and arbuscular mycorrhizal fungi: A soil microcosm study. Applied Soil Ecology. 2017;**119**:307-316

Section 2

Arbuscular Mycorrhizal Fungi

Chapter 3

Arbuscular Mycorrhizal Fungi in Intercropping Systems: Roles and Performance

Yunjian Xu and Fang Liu

Abstract

Arbuscular mycorrhizal fungi (AMF) have attracted significant interest in the field of sustainable agriculture. Intercropping is another sustainable practice improving the nutrient utilization efficiency. In an AMF-colonized intercropping system, intercropping has been found to increase the mycorrhization rate, including root colonization and spore population in the rhizosphere of plants. Root colonization of one plant by AMF is clearly influenced by their intercropping partners. Therefore, the selection of appropriate intercropping partners can be used to improve the activity of mycorrhizal symbiosis in crops. Furthermore, intercropping with different plant species can alter arbuscular mycorrhizal (AM) fungal diversity, and these different AM genera have distinct functions and benefits for plants in intercropping systems. Additionally, in certain intercropping systems, perennial plants serve as reservoirs of AMF inoculum for intercrops. In return, AM symbiosis enhances nutrient availability in the intercropping system, leading to positive effects of intercrops. Moreover, AMF exhibit bioprotective effects in intercropping systems, reducing the severity of plant diseases and/or compensating for plant biomass loss. However, these bioprotective effects depend on the intercropping partner rather than the degree of AM colonization. In conclusion, the combination of AMF benefits with intercropping holds great promise for improving nutrient utilization efficiency and plant health.

Keywords: arbuscular mycorrhizal fungi, intercropping system, symbiosis, nutrient, sustainable agriculture

1. Introduction

Arbuscular mycorrhizal fungi (AMF) are beneficial soil microorganisms that can form symbiosis with more than 80% of plants [1]. The beneficial functions of AMF in the field are important in sustainable agriculture. The formation of AM symbiosis is considered important in improving plant nutrition and resistance to plant diseases (especially soil-borne diseases) [2–4]. In mycorrhizal roots, nutrient and some chemical signal transfer are dependent on common mycorrhizal networks (CMN), which are formed by AMF hyphae enlarging the area for roots [5, 6] and linking the neighboring plants [7]. In addition, AMF have effects on different plant growth in a field by altering interspecific or intraspecific competitive situations [8, 9]. In return, the arbuscular mycorrhizal (AM) fungal community can be influenced by the host plants [10–12].

In a field, two or more crops are simultaneously cultivated called intercropping systems, and is receiving increasing global interest as an agricultural sustainable practice [13, 14]. Many crop combinations have been reported in intercropping systems, especially those cultivated with legume species as intercrops [15], such as sugarcane/soybean [16], maize/faba bean [17], maize/soybean [18, 19], and maize/peanut [19]. In the intercropping system, the intercrops were connected *via* CMN, in which nutrients and signaling compounds can be exchanged between the connected plants.

In intercropping systems, AMF symbiosis further improves plant growth by increasing nutrient uptake, improving soil condition, or improving disease resistance. In return, AMF colonization, diversity, and reservoir are related to interactions between different intercrops. The aim of this chapter is to review information on AMF function and its response to intercrops interactions in intercropping systems.

2. Benefits of AMF for plants

Arbuscular mycorrhizal fungi (AMF), belonging to the phylum Mucoromycota [20], can form mutualistic symbionts with more than 80% of land plants, including many agricultural crops [21]. A meta-analysis of 215 cases (involving 21 nurse species and 29 facilitated species) with/without AMF inoculation has shown that mycorrhizal fungi significantly improve plant facilitation mainly by increasing plant biomass (aboveground parts) and nutrient content [22]. AMF have a global impact on plant mineral nutrition, such as phosphorus (P) and nitrogen (N). AMF improves the uptake of mineral nutrients by increasing the surface area of absorbing and mobilizing sparsely available nutrients, which depends on AM fungal hyphae. In the root, the hyphae grow into root cortical cells, forming highly branched structures called arbuscules, which provide a large interface for nutrient exchange [1]. Mineral nutrients such as P and N are transported from AMF to plant hosts across the symbiotic interface [23]. In the soil, the extracellular hyphae form a fine network growing from colonized roots into the soil to expand the area for absorbing mineral nutrients from soil [24]. Mycorrhizal networks extend the absorbing surface area up to 40 times growing in every direction [25, 26], efficiently exploring the soil and increasing plant uptake of nutrients and water [27, 28]. This makes them essential for the establishment and productivity of plants in nutrient-limited soils and for the mitigation of plant stress [29]. Recent studies have shown that AMF also attracts other soil microorganisms helping them facilitate plant access to nutrients [30]. The intraradical AM fungal hyphae access carbon from the root and this fuels the growth of extraradical hyphae (ERH) in the surrounding soil, which feed on mineral nutrients essential for fungal growth and for delivery to the plant [31–33]. At the same time, growth of ERH increases the flow of carbon into the soil [34–36], which further recruits some special microorganisms to the rhizosphere.

Additionally, AMF increases plant resistance to biotic and abiotic stresses. For biotic stresses, in a biocontrol way, AMF plays a vital role in protecting the host from pathogens, which is one of the important strategies for ecologically sound management of plant diseases [37, 38]. For example, AMF inoculation suppresses tomato *Ralstonia* wilt and yield damage under the attack of *Ralstonia solanacearum* (soilborne pathogens), mainly through improved soil quality and alleviation of plant metabolic pressure [39]. AMF as a biocontrol for pepper *Phytophthora* blight has also been well reported [38, 40]. AMF *Funneliformis caledonium* significantly increased nutrient acquisition (N, P, and K), plant biomass, and pepper fruit yield under *Phytophthora capsici* inoculation [38]. Abiotic stress, such as drought, salinity, extreme temperatures, and pollution, present serious threats to agriculture. Fortunately, AMF have the ability to overcome adverse effects and improve plant performance in stressful environments. For example, AMF contributes to the production of low-Cd fruiting vegetables in Cd-contaminated fields: AMF symbiosis increase the P acquisition and biomass of cucumber plants, but reduce Cd concentrations in the plant by decreasing the efficiency of Cd acquisition by root [41].

3. Contributions of intercrops in intercropping systems

Intercropping is the cultivation of two or more crop species, or genotypes, in a field simultaneously, which is receiving increasing global interest as an agricultural practice [13]. Intercropping can be one of the ways to achieve "sustainable intensification" by achieving a real increase in yields without increasing inputs or by improving the stability of yields with fewer inputs [42]. For example, cereal/legume intercropping systems can reduce fertilizer inputs and have a positive impact on the environment [19, 43]. In legumes involved in intercropping cultivation, N fixed by legumes can be transferred to companion species [44], thus, the maize/soybean and maize/peanut intercropping systems show yield advantages compared to monocultures [19].

In addition to fixing N by legume, intercropping systems always increase yield and improve grain nutritional quality and ultimately human health, mainly due to intercropping contributing to increased nutrient (P, Fe, and Zn) uptake [45], which is associated with the rhizosphere processes. Intercropping may have a potential facilitating mechanism whereby some crop species can chemically mobilize unavailable forms of one or more limiting soil nutrients, such as P and micronutrients (iron (Fe), zinc (Zn), and manganese (Mn)). The P, Fe, and Zn-mobilizing crop species improve the nutrition of P, Fe, and Zn for themselves and neighboring non-mobilizing species with P, Fe, or Zn by releasing acid phosphatases, protons, and/or carboxylates, chelating substances in the rhizosphere, increasing the concentration of soluble inorganic P, Fe or Zn in the soil [46]. In cereal/legume intercropping, direct interspecific facilitation involves such processes, in which cereals increase Fe and Zn bioavailability while companion legumes benefit [45]. In terms of interspecific promotion, maize improves Fe nutrition in intercropped peanuts, faba beans improve N and P uptake by intercropped maize, and chickpea promotes P uptake by companion wheat [47]. Other intercropping systems, such as wheat/faba bean, which is also efficient in P utilization, but the reasons are not only due to positive interactions of rhizosphere processes but also due to complementary P uptake in the early growth stage [48].

A study of a maize and soybean intercropping system showed that root interactions are important to improve the soil microbiological environment, increase microbial populations and enzyme activities in the soil, and increase crop yields [49]. In addition to root interactions, studies in wheat/maize and wheat/soybean intercropping systems have shown that a positive impact of the border row and inner rows of the intercropped plants and more efficient use of soil nutrients than the monocrops result in both dominant and subordinate species in intercropped crops, which then produce higher yields than the corresponding wheat, maize, or soybean monocrops [47]. Additionally, intercropping improved plant resistance to some pathogens to improve yield. For example, intercropping with maize can provide relief to peppers against both wind-dispersed and soilborne pathogens [50]. Furthermore, intercropping systems largely impacted soil microbial functionalities and mycorrhizal soil infectivity, improving crop productivity [51]. These benefits resulting from intercrop should be closely related to changes recorded in soil microbial functionalities.

4. Advantages of the intercropping systems with AMF colonization in agriculture

4.1 Roles of AMF in agricultural intercropping system

AMF plays an important role in intercropping systems in two main ways: protecting plant health and improving plant growth and yield. Specifically, it involves the following: regulating the transfer, distribution, and acquisition of resources (e.g., nitrogen and phosphorus); influencing crop interactions; and resisting biotic (plant pathogens) and abiotic stresses (metals, drought) (**Figure 1**).

AMF play a vital role in optimizing resource use in the agricultural intercropping system, especially in resource-limited soils. Intercropping systems show a growth advantage for intercrop plants over those in monoculture systems [52, 53], while inoculation of AMF in intercropping systems, such as rice/mungbean, further increased the biomass of intercrop (mungbean) [54]. AMF can interconnect different individuals and even different plant species by forming a below-ground mycelial network from their hyphal branches [55, 56]. Such mycorrhizal networks can influence the distribution of nutrients among intercropped crops [57], and feedback in protecting plant health.

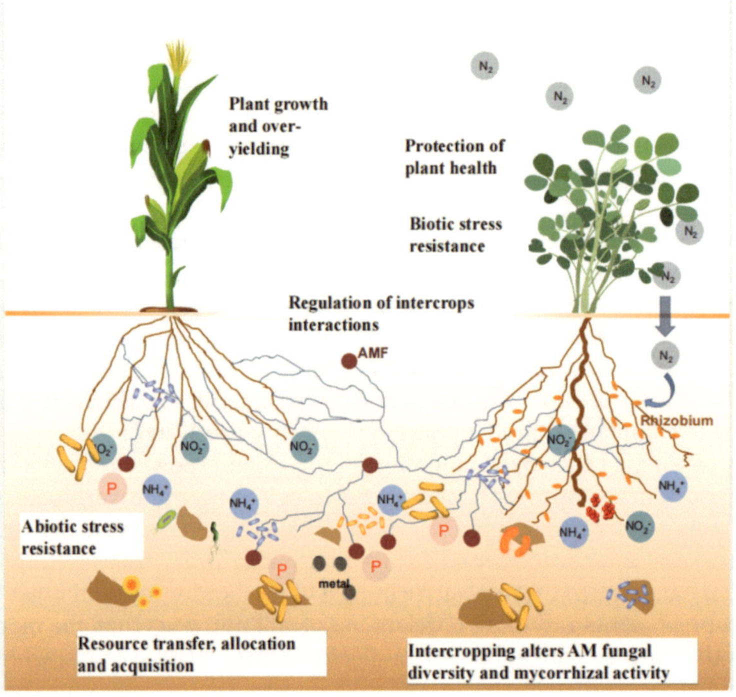

Figure 1.
Potential benefits of intercropping systems in symbiosis with arbuscular mycorrhizal fungi (AMF).

In low-input agriculture, AMF-inoculated cereal/legume intercropping typically provides over-yielding. Mycorrhizal networks contribute to the over-yielding in intercropping by improving the nutrient uptake of mycorrhiza-dependent plant species. Fungal hyphal networks AM could transfer nutrients among intercrops, increasing plant P uptake, and when the AMF was absent in intercropping eggplant (*Solanum melongena* L.) with maize, the expected advantages disappeared [58]. In a maize-grass pea intercropping system, inoculation with AMF resulted in mutual reinforcement of maize and grass pea under low P and low water conditions, but as the availability of resources increased, maize was facilitated, but grass pea had a neutral role or facilitator [59]. In the wheat and faba bean intercropping system, the inoculation of AMF *Funneliformis mosseae* enhances the complementary resource for intercrops under N and P limiting conditions [60]. In addition, AMF show different responses to different forms of P [$Ca_{10}(OH)_2(PO_4)_6$ (Ca-P), sodium phytate (Na-P) and KH_2PO_4 (K-P)] in coix (*Coix lacryma-jobi* L.)/faba bean (*Vicia faba* L.) intercropping systems for the intercrops, indicated the differences in the strategies utilized in the forms P of coix and faba bean, which improve intercropping productivity [61]. Except for nutrients, AMF inoculation alleviates the negative effects of drought conditions on finger millet in finger millet/pigeon pea intercropping systems [62].

AMF play an important role in mediating interactions among neighboring plants. In intercropping systems, mycorrhizal networks showed different effects on intraspecific and interspecific interactions. In the presence of AMF, maize and faba bean benefited from mycorrhizal networks and intermingling root systems, but weeds (foxtail) were inhibited by interspecific interactions [63]. AMF can change the competitive relationships between intercrops by modifying plants' nutritional levels [64]. In the maize/chili pepper intercropping system, AMF form a better symbiotic relationship with maize, which also supply part of photosynthetic C to increase AM fungal propagules in the chili pepper compartment, increasing the competitive capacity of P of pepper against maize and increasing pepper fruit yield [57]. In the faba bean/wheat intercropping system, the colonization of AMF favored the stronger competitor in the mixture (wheat) without negatively affecting the companion species (faba bean), including positive effects on root biomass, specific root length, root density, and uptake of P, Fe, and Zn [17]. Additionally, AMF colonization and AMF species can also affect the competitive relationships between crops and weeds [65]. The colonization of the AM root of tomato was clearly affected by its intercropping partners: The tomato intercropped with leek showed a higher AM colonization rate than intercropped with itself [66].

4.2 Resource transfer between intercrops through CMNs

AMF extend the root surface by more than 10 cm per gram of soil [67], and such mycelial network plays a significant role in mediating the flow of mineral nutrients from the soil to the host plants. These mycelial networks can be supported by several host plants of either the same or even different plant species, thereby forming common mycorrhizal networks (CMN) [68]. The formation of CMN affects the distribution and transfer of resources from a plant to other plants, such as nutrients (N, P, C) [69, 70] and defense signals [71, 72], which may affect plant competitive outcomes and community composition through differential resource allocation. The study has shown that flax acquired 80 and 94% of total N and P by CMN, respectively, while neighboring sorghum (*Sorghum bicolor* L.) acquired 20 and 6% of N and P by the CMN [73]. N transfer through direct and indirect pathways in cereal/legume intercropping systems [74, 75]. The direct pathway is N fixed in legumes transferred to related cereal plants through CMN, and the indirect pathway is that legume residues and root exudates release N into the rhizosphere, which is then absorbed by the intercropped cereal or mycorrhizal hyphae. In non-mycorrhizal roots, N was only transferred by the root network, while AMF symbiosis can provide a mycorrhizal pathway for N transferred by the hyphae network, which increases contact between plants and contact with roots, increasing N transport [76]. In the faba bean/wheat intercropping system, the AMF symbiosis increased N transfer from the faba bean to wheat by 20% [17].

In addition to N and P, CMN can mediate K trade to benefit individuals in intercropping system, but maybe there exists an asymmetry between intercrops. In the AMF inoculated tomato/potato-onion system, intercropping increased the dry weight of the plant and the K content of tomato, but decreased those of potato-onion, because K transfer between plants has a strong asymmetry: more K was transferred from potato-onion to tomato than from tomato to potato-onion [73]. Reasons for asymmetry in nutrient gains between plants have been studied in the cereal/legume intercropping system. They reveal the asymmetry in nutrient gains by a mixture of millet (*Setaria italica* L.) and chickpea (*Cicer arietinum* L.), dependent on which plant species was the donor or receiver: chickpea inoculation increased the acquisition of N and P and the biomass of both chickpea (donor) and millet (receiver), leading to overyielding of the mixture [77]. Therefore, they suggest that the introduction of mycorrhizal networks by legumes in intercropping could be an important factor contributing to the magnitude of the intercropping effect. It's worth noting that different AMF show different contributions to plant growth and nutrient uptake through CMN. For example, the CMN formed by *Funneliformis mosseae* was greater than those of *F. intraradices* [78].

The establishment of CMNs appears to influence intraspecific and interspecific interactions between donor and receiver plants [78]. Mycorrhization decreased the proportion of N transferred from pea to wheat but increased the proportion transferred to flax and chicory, suggesting that AM symbiosis influences the distribution of N, and thus affects competition between neighboring plants of different species [70]. In *Arctic tundra*, belowground carbon (C) transfer by CMNs is enough to potentially alter plant interactions, increasing the competitive ability and monodominance of dwarf birch (*Betula nana* L.) [64]. The different plant-plant interactions that result from the symbiosis of AMF will provide different outcomes based on the two processes: the diversity of resources provided by AMF to its host (resource dissimilarity) and different ways of allocating these resources (resource allocation) [79].

4.3 Effects of intercropping on AMF colonization

Intercropping has an impact on the community and the functions of AMF in soils in many aspects (**Figure 1**). Alter AM fungal diversity. Conventional agricultural practices have been shown to have a negative impact on the abundance and diversity of AM fungi [12], but intercropping has shown the potential to reduce the negative impact. Increased plant diversity can alter the diversity of species of fungi [80]. In intercropping systems, such as maize/soybeans and oat/pea, the AMF diversity was higher in an intercropped system compared to their respective monocropping system [11, 81]. Tree-based intercropping systems support a more abundant and diverse AM fungal community compared to conventionally managed systems [12]. However, some studies intend to show zero or negative significant effects on the fungal community of AMF due to the incorporation of trees into agricultural systems, which may be a function of different cultivation techniques, climatic variation, or various tree-crop combinations used within different tree-based intercropping systems [82–84]. Furthermore, the amount of growth of fungal species of AMF also depends on the associated host plant species [85, 86]. In intercropping systems, such as the maize/soybean intercropping system, different N applications also change the AMF diversity and community: Glomus_f_Glomeraceae in maize soil and roots increased significantly with increasing N application levels, but decreased with increasing N application levels in soybean soil and roots [18]. In the maize/soybean strip intercropping systems, both N application and the cropping pattern affect the diversity of the AM fungal community, but the N application had a stronger influence than the cropping pattern: Lower N application shows significantly higher diversity and co-occurring species than higher N application [87].

Improve mycorrhizal symbiosis activity. Intercropping can improve AM symbiosis activity among intercrops, such as increasing mycorrhization, including root colonization and spore population in the rhizosphere. In intercropping systems of sorghum/ana tree (*Faidherbia albida)* and soybean/peacock flower (*Albizia gummifera*) or broad-leaved croton (*Croton macrostachyus*), both root colonization and soil sporulation of AMF are increased [88, 89]. Increases in mycorrhizal colonization in intercropped systems consistent with more than two plant species have also been found in groundnuts and sweet potato as intercrops of banana [90], wheat and maize intercropped with faba bean [91]. In the maize/soybean intercropping system, AMF-secreted glomalin-related soil protein (GRSP) concentrations are generally increased in the soil of the rhizosphere of the crop compared to monoculture [81]. Intercropping promotes the release of a wide range of organic compounds that contribute to the activity of the soil microbiota, including AMF, increasing the colonization and species diversity of these fungi [80]. However, the study also finds that AMF colonization of tomato is increased, not changed, or decreased when intercropped with leek, cucumber/basil, and fennel, respectively [66], which indicated a neighbor effect on plant-AM fungal interactions [92].

Perennial plants severed as reservoirs of inoculum in intercropping systems. In the tree-based intercropping system (Agroforestry), perennial trees play a role in maintaining active populations and mycelial networks of AMF [12]. These trees always acted as reservoirs for the AMF inoculum for the intercrops. AMF are always nonspecific in their selection of host [93], and thus the AMF soil habitat could serve as potential inoculants for the implantation of plant species [94]. Alternatively, native AMF which undergo a variety of modifications to survive in their habitats can be isolated and used as potential inoculants for inoculation into plant species of interest [95].

Improvement of disease resistance in host plants in intercropping systems. Continuous monoculture of pepper often leads to increases in soil pathogen populations [96], especially the most economically destructive soilborne diseases. *Phytophthora* blight causes severe plant disease and thus enormous yield loss [97]. By elevating root mycorrhization, intercropping with maize could suppress pepper *Phytophthora* blight and increase fruit biomass [98]. However, not all intercropping plants have a positive effect on disease resistance after AMF colonization. With different intercropping partners (leek, cucumber, basil, fennel, or tomato itself), only in tomato/leek and tomato/basil combinations, AMF inoculation reduces the negative effects of *F. oxysporum* f. sp. *lycopersici* on tomato biomass [66]. Therefore, the bioprotective effect of AMF resulting in reduced disease severity of *F. oxysporum* f. sp. *lycopersici* and/or compensation of plant biomass does not depend on the extent of AM colonization, but more on the intercropping partner [66].

5. Conclusions

Combining arbuscular mycorrhizal fungi (AMF) and intercropping can improve resource utilization efficiency in the field of sustainable agriculture. In the AMF-colonized intercropping system, the root colonization of one plant by AMF is influenced by its intercrops. Furthermore, different intercropping species can alter the arbuscular mycorrhizal (AM) fungal community. Moreover, the bioprotective effects of AMF depend on the intercropping partner rather than the degree of AM colonization. Thus, in agricultural practice, the application of intercropping in combination with AMF needs to be tested for each intercropping arrangement separately, and the appropriate selection of intercropping partners can be beneficial for each intercrop. Additionally, the characterization of native AMF communities and their application to intercropping systems may be a useful tool for agricultural development.

Acknowledgements

This work was supported by the National Natural Science Foundation of China (31902104, 32101358), the Yunnan Fundamental Research Projects (202301AT070106), the Double Top University Fund of Yunnan University and the Yunnan Revitalization Talent Support Program.

Conflict of interest

The authors declare no conflict of interest.

Statements and declarations

The authors declare that the research was conducted in the absence of commercial or financial relationships that could be construed as a potential conflict of interest.

Author details

Yunjian Xu[1]* and Fang Liu[2]

1 Ministry of Education Key Laboratory for Transboundary Ecosecurity of Southwest China, Yunnan Key Laboratory of Plant Reproductive Adaptation and Evolutionary Ecology and Centre for Invasion Biology, Institute of Biodiversity, School of Ecology and Environmental Science, Yunnan University, Kunming, Yunnan, China

2 School of Agriculture, Yunnan University, Kunming, China

*Address all correspondence to: xuyunjian1992@ynu.edu.cn

References

[1] Smith SE, Read DJ. Mycorrhizal Symbiosis. London: Academic Press; 2008

[2] Wang W, Shi J, Xie Q, Jiang Y, Yu N, Wang E. Nutrient exchange and regulation in arbuscular mycorrhizal symbiosis. Molecular Plant. 2017;**10**(9):1147-1158

[3] Azcón-Aguilar C, Jaizme-Vega MC, Calvet C. The contribution of arbuscular mycorrhizal fungi to the control of soil-borne plant pathogens. In: Gianinazzi S, Schüepp H, Barea JM, Haselwandter K, editors. Mycorrhizal Technology in Agriculture: From Genes to Bioproducts. Basel: Birkhäuser Basel; 2002. pp. 187-197

[4] Whipps JM. Prospects and limitations for mycorrhizas in biocontrol of root pathogens. Canadian Journal of Botany. 2004;**82**(8):1198-1227

[5] Zhang L, Zhou J, George TS, Limpens E, Feng G. Arbuscular mycorrhizal fungi conducting the hyphosphere bacterial orchestra. Trends in Plant Science. 2022;**27**(4):402-411

[6] Homulle Z, George TS, Karley AJ. Root traits with team benefits: Understanding belowground interactions in intercropping systems. Plant and Soil. 2022;**471**(1):1-26

[7] Wipf D, Krajinski F, van Tuinen D, Recorbet G, Courty P-E. Trading on the arbuscular mycorrhiza market: From arbuscules to common mycorrhizal networks. New Phytologist. 2019;**223**(3):1127-1142

[8] Van Der Heijden MGA, Wiemken A, Sanders IR. Different arbuscular mycorrhizal fungi alter coexistence and resource distribution between co-occurring plant. New Phytologist. 2003;**157**(3):569-578

[9] Schroeder-Moreno MS, Janos DP. Intra- and inter-specific density affects plant growth responses to arbuscular mycorrhizas. Botany. 2008;**86**(10):1180-1193

[10] Bever JD. Host-specificity of AM fungal population growth rates can generate feedback on plant growth. Plant and Soil. 2002;**244**(1):281-290

[11] Lee A, Neuberger P, Omokanye A, Hernandez-Ramirez G, Kim K, Gorzelak MA. Arbuscular mycorrhizal fungi in oat-pea intercropping. Scientific Reports. 2023;**13**(1):390

[12] Bainard LD, Klironomos JN, Gordon AM. Arbuscular mycorrhizal fungi in tree-based intercropping systems: A review of their abundance and diversity. Pedobiologia. 2011;**54**(2):57-61

[13] Mazzafera P, Favarin JL, Andrade SAL. Editorial: Intercropping systems in sustainable agriculture. Frontiers in Sustainable Food Systems. 2021;**5**

[14] Glaze-Corcoran S, Hashemi M, Sadeghpour A, Jahanzad E, Keshavarz Afshar R, Liu X, et al. Chapter five - understanding intercropping to improve agricultural resiliency and environmental sustainability. In: Sparks DL, editor. Advances in Agronomy. Vol. 162. Academic Press Inc, Elsevier Science; 2020. pp. 199-256

[15] Stagnari F, Maggio A, Galieni A, Pisante M. Multiple benefits of legumes for agriculture sustainability: An overview. Chemical and Biological Technologies in Agriculture. 2017;**4**(1):2

[16] Luo S, Yu L, Liu Y, Zhang Y, Yang W, Li Z, et al. Effects of reduced nitrogen input on productivity and N_2O emissions in a sugarcane/soybean intercropping system. European Journal of Agronomy. 2016;**81**:78-85

[17] Ingraffia R, Amato G, Alfonso SF, Dario G. Impacts of arbuscular mycorrhizal fungi on nutrient uptake, N_2 fixation, N transfer, and growth in a wheat/faba bean intercropping system. PLoS One. 2019;**14**(3):e0213672

[18] Zhang R, Mu Y, Li X, Li S, Sang P, Wang X, et al. Response of the arbuscular mycorrhizal fungi diversity and community in maize and soybean rhizosphere soil and roots to intercropping systems with different nitrogen application rates. Science of The Total Environment. 2020;**740**:139810

[19] Zhang WP, Gao SN, Li ZX, Xu HS, Yang H, Yang X, et al. Shifts from complementarity to selection effects maintain high productivity in maize/legume intercropping systems. Journal of Applied Ecology. 2021;**58**(11):2603-2613

[20] Spatafora JW, Chang Y, Benny GL, Lazarus K, Smith ME, Berbee ML, et al. A phylum-level phylogenetic classification of zygomycete fungi based on genome-scale data. Mycologia. 2016;**108**(5):1028-1046

[21] Brundrett MC, Tedersoo L. Evolutionary history of mycorrhizal symbioses and global host plant diversity. New Phytologist. 2018;**220**(4):1108-1115

[22] Montesinos-Navarro A, Valiente-Banuet A, Verdú M. Mycorrhizal symbiosis increases the benefits of plant facilitative interactions. Ecography. 2019;**42**(3):447-455

[23] Bonfante P, Genre A. Mechanisms underlying beneficial plant–fungus interactions in mycorrhizal symbiosis. Nature Communications. 2010;**1**(1):48

[24] Mosse B. Observations on the extra-matrical mycelium of a vesicular-arbuscular endophyte. Transactions of the British Mycological Society. 1959;**42**(4):439-448

[25] Giovannetti M, Fortuna P, Citernesi AS, Morini S, Nuti MP. The occurrence of anastomosis formation and nuclear exchange in intact arbuscular mycorrhizal networks. New Phytologist. 2001;**151**(3):717-724

[26] Pepe A, Giovannetti M, Sbrana C. Different levels of hyphal self-incompatibility modulate interconnectedness of mycorrhizal networks in three arbuscular mycorrhizal fungi within the Glomeraceae. Mycorrhiza. 2016;**26**(4):325-332

[27] Hodge A, Campbell CD, Fitter AH. An arbuscular mycorrhizal fungus accelerates decomposition and acquires nitrogen directly from organic material. Nature. 2001;**413**(6853):297-299

[28] Clark RB, Zeto SK. Mineral acquisition by arbuscular mycorrhizal plants. Journal of Plant Nutrition. 2000;**23**(7):867-902

[29] Kakouridis A, Hagen JA, Kan MP, Mambelli S, Feldman LJ, Herman DJ, et al. Routes to roots: Direct evidence of water transport by arbuscular mycorrhizal fungi to host plants. New Phytologist. 2022;**236**(1):210-221

[30] Emmett BD, Lévesque-Tremblay V, Harrison MJ. Conserved and reproducible bacterial communities associate with extraradical hyphae of arbuscular mycorrhizal fungi. The ISME Journal. 2021;**15**(8):2276-2288

[31] Pearson JN, Jakobsen I. The relative contribution of hyphae and roots to

phosphorus uptake by arbuscular mycorrhizal plants, measured by dual labelling with ^{32}P and ^{33}P. New Phytologist. 1993;**124**(3):489-494

[32] Hodge A, Fitter AH. Substantial nitrogen acquisition by arbuscular mycorrhizal fungi from organic material has implications for N cycling. Proceedings of the National Academy of Sciences. 2010;**107**(31):13754-13759

[33] Wang S, Chen A, Xie K, Yang X, Luo Z, Chen J, et al. Functional analysis of the OsNPF4.5 nitrate transporter reveals a conserved mycorrhizal pathway of nitrogen acquisition in plants. Proceedings of the National Academy of Sciences. 2020;**117**(28):16649-16659

[34] Johnson D, Leake JR, Ostle N, Ineson P, Read DJ. In situ$^{13}CO_2$ pulse-labelling of upland grassland demonstrates a rapid pathway of carbon flux from arbuscular mycorrhizal mycelia to the soil. New Phytologist. 2002;**153**(2):327-334

[35] Finlay RD. Ecological aspects of mycorrhizal symbiosis: With special emphasis on the functional diversity of interactions involving the extraradical mycelium. Journal of Experimental Botany. 2008;**59**(5):1115-1126

[36] Kaiser C, Kilburn MR, Clode PL, Fuchslueger L, Koranda M, Cliff JB, et al. Exploring the transfer of recent plant photosynthates to soil microbes: Mycorrhizal pathway vs direct root exudation. New Phytologist. 2015;**205**(4):1537-1551

[37] Hou S, Zhang Y, Li M, Liu H, Wu F, Hu J, et al. Concomitant biocontrol of pepper Phytophthora blight by soil indigenous arbuscular mycorrhizal fungi via upfront film-mulching with reductive fertilizer and tobacco waste. Journal of Soils and Sediments. 2020;**20**(1):452-460

[38] Hu J, Hou S, Li M, Wang J, Wu F, Lin X. The better suppression of pepper Phytophthora blight by arbuscular mycorrhizal (AM) fungus than *Purpureocillium lilacinum* alone or combined with AM fungus. Journal of Soils and Sediments. 2020;**20**(2):792-800

[39] Li M, Hou S, Wang J, Hu J, Lin X. Arbuscular mycorrhizal fungus suppresses tomato (*Solanum lycopersicum* Mill.) *Ralstonia* wilt via establishing a soil–plant integrated defense system. Journal of Soils and Sediments. 2021;**21**(11):3607-3619

[40] Reyes Tena A, Rincon-Enriquez G, Lopez-Perez L, Quiñones Aguilar EE. Effect of mycorrhizae and actinomycetes on growth and bioprotection of *Capsicum annuum* L. against *Phytophthora capsici*. Pakistan Journal of Agricultural Sciences. 2017;**54**:513-522

[41] Hu J, Tsang W, Wu F, Wu S, Wang J, Lin X, et al. Arbuscular mycorrhizal fungi optimize the acquisition and translocation of Cd and P by cucumber (*Cucumis sativus* L.) plant cultivated on a Cd-contaminated soil. Journal of Soils and Sediments. 2016;**16**(9):2195-2202

[42] Brooker RW, Bennett AE, Cong W-F, Daniell TJ, George TS, Hallett PD, et al. Improving intercropping: A synthesis of research in agronomy, plant physiology and ecology. New Phytologist. 2015;**206**(1):107-117

[43] Li XF, Wang ZG, Bao XG, Sun JH, Yang SC, Wang P, et al. Long-term increased grain yield and soil fertility from intercropping. Nature Sustainability. 2021;**4**(11):943-950

[44] Moyer-Henry KA, Burton JW, Israel DW, Rufty TW. Nitrogen transfer between plants: A ^{15}N natural abundance study with crop and weed species. Plant and Soil. 2006;**282**(1):7-20

[45] Xue Y, Xia H, Christie P, Zhang Z, Li L, Tang C. Crop acquisition of phosphorus, iron and zinc from soil in cereal/legume intercropping systems: A critical review. Annals of Botany. 2016;**117**(3):363-377

[46] Li L, Tilman D, Lambers H, Zhang FS. Plant diversity and overyielding: Insights from belowground facilitation of intercropping in agriculture. New Phytologist. 2014;**203**(1):63-69

[47] Zhang F, Li L. Using competitive and facilitative interactions in intercropping systems enhances crop productivity and nutrient-use efficiency. Plant and Soil. 2003;**248**(1):305-312

[48] Li C, Dong Y, Li H, Shen J, Zhang F. Shift from complementarity to facilitation on P uptake by intercropped wheat neighboring with faba bean when available soil P is depleted. Scientific Reports. 2016;**6**(1):18663

[49] Zhang X, Huang G, Bian X, Zhao Q. Effects of nitrogen fertilization and root interaction on the agronomic traits of intercropped maize, and the quantity of microorganisms and activity of enzymes in the rhizosphere. Plant and Soil. 2013;**368**(1):407-417

[50] Yang M, Zhang Y, Qi L, Mei X, Liao J, Ding X, et al. Plant-plant-microbe mechanisms involved in soil-borne disease suppression on a maize and pepper intercropping system. PLoS One. 2015;**9**(12):e115052

[51] Wahbi S, Prin Y, Thioulouse J, Sanguin H, Baudoin E, Maghraoui T, et al. Impact of wheat/faba bean mixed cropping or rotation systems on soil microbial functionalities. Frontiers in Plant Science. 2016;**7**

[52] Hauggaard-Nielsen H, Jensen ES. Facilitative root interactions in intercrops. Plant and Soil. 2005;**274**(1):237-250

[53] Chapagain T, Riseman A. Barley–pea intercropping: Effects on land productivity, carbon and nitrogen transformations. Field Crops Research. 2014;**166**:18-25

[54] Xiao T, Yang Q, Ran W, Xu G, Shen Q. Effect of inoculation with arbuscular mycorrhizal fungus on nitrogen and phosphorus utilization in upland rice-mungbean intercropping system. Agricultural Sciences in China. 2010;**9**(4):528-535

[55] Giovannetti M, Sbrana C, Avio L, Strani P. Patterns of below-ground plant interconnections established by means of arbuscular mycorrhizal networks. New Phytologist. 2004;**164**(1):175-181

[56] Ren L, Zhang N, Wu P, Huo H, Xu G, Wu G. Arbuscular mycorrhizal colonization alleviates *fusarium* wilt in watermelon and modulates the composition of root exudates. Plant Growth Regulation. 2015;**77**(1):77-85

[57] Hu J, Li M, Liu H, Zhao Q, Lin X. Intercropping with sweet maize (*Zea mays* L. var. rugosa Bonaf.) expands P acquisition channels of chili pepper (*Capsicum annuum* L.) via arbuscular mycorrhizal hyphal networks. Journal of Soils and Sediments. 2019;**19**(4):1632-1639

[58] Vandermeer JH. The Ecology of Intercropping. Cambridge: Cambridge University Press; 1989

[59] Zhu SG, Duan HX, Tao HY, Zhu L, Zhou R, Yang YM, et al. Arbuscular mycorrhiza changes plant facilitation patterns and increases resource use efficiency in intercropped annual plants. Applied Soil Ecology. 2023;**191**:105030

[60] Qiao X, Bei S, Li C, Dong Y, Li H, Christie P, et al. Enhancement of faba bean competitive ability by arbuscular mycorrhizal fungi is highly correlated with dynamic nutrient acquisition by competing wheat. Scientific Reports. 2015;**5**(1):8122

[61] Bei S, Xu M, Lyu X, Chen C, Li A, Qiao X. Arbuscular mycorrhizal fungi enhanced coix responses to phosphorous forms but not for faba bean in intercropping systems, under controlled environment. Agronomy Journal. 2021;**113**(3):2578-2590

[62] Saharan Jakhar K, Schütz L, Kahmen A, Wiemken A, Boller T, Mathimaran N. Finger millet growth and nutrient uptake is improved in intercropping with pigeon pea through "biofertilization" and "bioirrigation" mediated by arbuscular mycorrhizal fungi and plant growth promoting rhizobacteria. Frontiers in Environmental Science. 2018;**6**:46

[63] Qiao X, Bei S, Li H, Christie P, Zhang F, Zhang J. Arbuscular mycorrhizal fungi contribute to overyielding by enhancing crop biomass while suppressing weed biomass in intercropping systems. Plant and Soil. 2016;**406**(1):173-185

[64] Deslippe JR, Simard SW. Below-ground carbon transfer among Betula nana may increase with warming in *Arctic tundra*. New Phytologist. 2011;**192**(3):689-698

[65] Rashidi S, Yousefi AR, Pouryousef M, Goicoechea N. Mycorrhizal impact on competitive relationships and yield parameters in *Phaseolus vulgaris* L. — Weed mixtures. Mycorrhiza. 2021;**31**(5):599-612

[66] Hage-Ahmed K, Krammer J, Steinkellner S. The intercropping partner affects arbuscular mycorrhizal fungi and *Fusarium oxysporum* f. sp. *lycopersici* interactions in tomato. Mycorrhiza. 2013;**23**(7):543-550

[67] Smith SE, Smith FA. Roles of arbuscular mycorrhizas in plant nutrition and growth: New paradigms from cellular to ecosystem scales. Annual Review of Plant Biology. 2011;**62**(1):227-250

[68] Van Der Heijden MGA, Horton TR. Socialism in soil? The importance of mycorrhizal fungal networks for facilitation in natural ecosystems. Journal of Ecology. 2009;**97**(6):1139-1150

[69] Luo X, Liu Y, Li S, He X. Interplant carbon and nitrogen transfers mediated by common arbuscular mycorrhizal networks: Beneficial pathways for system functionality. Frontiers in Plant Science. 2023;**14**

[70] Ingraffia R, Giambalvo D, Frenda AS, Roma E, Ruisi P, Amato G. Mycorrhizae differentially influence the transfer of nitrogen among associated plants and their competitive relationships. Applied Soil Ecology. 2021;**168**:104127

[71] Babikova Z, Gilbert L, Bruce TJA, Birkett M, Caulfield JC, Woodcock C, et al. Underground signals carried through common mycelial networks warn neighbouring plants of aphid attack. Ecology Letters. 2013;**16**(7):835-843

[72] Song YY, Ye M, Li C, He X, Zhu-Salzman K, Wang RL, et al. Hijacking common mycorrhizal networks for herbivore-induced defence signal transfer between tomato plants. Scientific Reports. 2014;**4**(1):3915

[73] Walder F, Niemann H, Natarajan M, Lehmann MF, Boller T, Wiemken A. Mycorrhizal networks: Common goods

of plants shared under unequal terms of trade. Plant Physiology. 2012;**159**(2):789-797

[74] He X, Critchley C, Ng H, Bledsoe C. Nodulated N_2-fixing *Casuarina cunninghamiana* is the sink for net N transfer from non-N_2-fixing *Eucalyptus maculata* via an ectomycorrhizal fungus *Pisolithus* sp. using $^{15}NH_4^+$ or $^{15}NO_3^-$ supplied as ammonium nitrate. The New Phytologist. 2005;**167**(3):897-912

[75] Xiao Y, Li L, Zhang F. Effect of root contact on interspecific competition and N transfer between wheat and fababean using direct and indirect ^{15}N techniques. Plant and Soil. 2004;**262**:45-54

[76] Chu G, Shen Q, Li Y, Zhang J, Wang S. Researches on Bi-directional N transfer between the intercropping system of groundnut with rice cultivated in aerobic soil using ~(15)N foliar labelling method. Acta Ecologica Sinica. 2004;**24**(2):278-284

[77] Li C, Li H, Hoffland E, Zhang F, Zhang J, Kuyper TW. Common mycorrhizal networks asymmetrically improve chickpea N and P acquisition and cause overyielding by a millet/chickpea mixture. Plant and Soil. 2022;**472**(1):279-293

[78] Qiao X, Guo X, Li A. Common mycorrhizal networks contribute to overyielding in faba bean/coix intercropping systems. Agronomy Journal. 2020;**112**(4):2598-2607

[79] Montesinos-Navarro A, Valiente-Banuet A, Verdú M. Processes underlying the effect of mycorrhizal symbiosis on plant-plant interactions. Fungal Ecology. 2019;**40**:98-106

[80] Pires GC, Eloá de Lima M, Zanchi CS, Moretti de Freitas C, Andrade de Souza JM, Andrea de Camargo T, et al. Arbuscular mycorrhizal fungi in the rhizosphere of soybean in integrated crop livestock systems with intercropping in the pasture phase. Rhizosphere. 2021;**17**:100270

[81] Lu M, Zhao J, Lu Z, Li M, Yang J, Fullen M, et al. Maize–soybean intercropping increases soil nutrient availability and aggregate stability. Plant and Soil. 2023

[82] Chirko CP, Gold MA, Nguyen PV, Jiang JP. Influence of direction and distance from trees on wheat yield and photosynthetic photon flux density (Qp) in a Paulownia and wheat intercropping system. Forest Ecology and Management. 1996;**83**(3):171-180

[83] Jose S, Gillespie AR, Pallardy SG. Interspecific interactions in temperate agroforestry. Agroforestry Systems. 2004;**61**(1):237-255

[84] Jose S, Gillespie AR, Seifert JR, Biehle DJ. Defining competition vectors in a temperate alley cropping system in the midwestern USA: 2. Competition for water. Agroforestry Systems. 2000;**48**(1):41-59

[85] Bever JD, Morton JB, Antonovics J, Schultz PA. Host-dependent sporulation and species diversity of arbuscular mycorrhizal fungi in a mown grassland. Journal of Ecology. 1996;**84**(1):71-82

[86] Eom AH, Hartnett DC, Wilson GWT. Host plant species effects on arbuscular mycorrhizal fungal communities in tallgrass prairie. Oecologia. 2000;**122**(3):435-444

[87] Zeng H, Yu L, Liu P, Wang Z, Chen Y, Wang J. Nitrogen fertilization has a stronger influence than cropping pattern on AMF community in maize/soybean strip intercropping systems. Applied Soil Ecology. 2021;**167**:104034

[88] Birhane E, Gebremeskel K, Taddesse T, Hailemariam M, Hadgu KM, Norgrove L, et al. Integrating Faidherbia albida trees into a sorghum field reduces striga infestation and improves mycorrhiza spore density and colonization. Agroforestry Systems. 2018;**92**(3):643-653

[89] Hailemariam M, Birhane E, Asfaw Z, Zewdie S. Arbuscular mycorrhizal association of indigenous agroforestry tree species and their infective potential with maize in the rift valley, Ethiopia. Agroforestry Systems. 2013;**87**(6):1261-1272

[90] Mandou MS, Adamou S, Nwaga D, Etoa R-X. Intercrops influence mycorrhizal symbiosis development, growth and nutrient uptake of banana. International Journal of Current Microbiology and Applied Sciences. 2016;**5**:84-94

[91] Wahbi S, Maghraoui T, Hafidi M, Sanguin H, Oufdou K, Prin Y, et al. Enhanced transfer of biologically fixed N from faba bean to intercropped wheat through mycorrhizal symbiosis. Applied Soil Ecology. 2016;**107**:91-98

[92] Li M, Hu J, Lin X. The roles and performance of arbuscular mycorrhizal fungi in intercropping systems. Soil Ecology Letters. 2022;**4**(4):319-327

[93] Hoeksema J. Investigating the disparity in host specificity between AM and EM fungi: Lessons from theory and better studied systems. Oikos. 1999;**84**:327

[94] Fernández Di Pardo A, Mancini M, Cravero V, Gil-Cardeza ML. Diagnose of indigenous arbuscular mycorrhizal communities associated to *Cynara cardunculus* L. var. *altilis* and var. *sylvestris*. Current Microbiology. 2021;**78**(1):190-197

[95] Shukla A, Kumar A, Jha A, Dhyani SK, Vyas D. Cumulative effects of tree-based intercropping on arbuscular mycorrhizal fungi. Biology and Fertility of Soils. 2012;**48**(8):899-909

[96] Khan M, Cheng Z, Khan A, Rana S, Ghazanfar B. Pepper-garlic intercropping system improves soil biology and nutrient status in plastic tunnel. International Journal of Agriculture and Biology. 2015;**17**:869-880

[97] Li Q, Ai G, Shen D, Zou F, Wang J, Bai T, et al. A *Phytophthora capsici* effector targets ACD11 binding partners that regulate ROS-mediated defense response in Arabidopsis. Molecular Plant. 2019;**12**(4):565-581

[98] Hu J, Hou S, Cai P, Li M, Cheng Z, Wu F, et al. Intercropping with maize (*Zea mays* L.) enhanced the suppression of pepper (*Capsicum annuum* L.) phytophthora blight by arbuscular mycorrhizal fungi. Journal of Soils and Sediments. 2023;**23**(2):891-901

Chapter 4

Diversity and Function of Mycorrhizal Fungi

YingWu Shi, XinXiang Niu, Hongmei Yang, Ming Chu, Huifang Bao, Ning Wang, Faqiang Zhan, Xuanqi Long, Rong Yang, Qing Lin and Kai Lou

Abstract

With the progress of science and technology and the deepening of people's understanding of mycorrhizal fungi, the diversity and function of mycorrhizal fungi have attracted attention of scholars, and the research on mycorrhizal application technology has been strengthened. In order to grasp the latest progress and current situation of mycorrhizal fungi diversity research, and clarify the achievements in the research and application fields of mycorrhizal fungi diversity and function, this study summarizes the latest research progress of mycorrhizal fungi diversity and function, which are discussed. The morphological characteristics, anatomical characteristics, host plant species and mycorrhizal fungal types, the species and genetic diversity of arbuscular mycorrhizal fungi, the origin of arbuscular mycorrhizal fungi, and the influencing factors of arbuscular mycorrhizal fungal diversity are discussed. A lot of work has been done on the species, geographical distribution, ecological characteristics, and resource investigation of ectomycorrhizal fungi. More and more ECM fungal resources have been detected and identified. The ecological function of mycorrhizal fungi is manifested in the aspects of plant community and plant ecosystem stability by improving ecosystem productivity. Mycorrhizal fungi can form symbionts with plants, enter the food web as food, and affect terrestrial ecosystems.

Keywords: mycorrhizal fungi, arbuscular mycorrhizal, ectomycorrhizal, diversity, function

1. Introduction

Mycorrhiza is a symbiont formed by plant roots and some fungi in soil. It is widely found in terrestrial ecosystems, including forests, fields, and grasslands. Mycorrhiza plays an important role in the ecosystem [1–3]. More than 80% of terrestrial ecosystems can form mycorrhiza with mycorrhizal fungi. This symbiosis is beneficial to both host plants and fungi. According to the morphological characteristics, anatomical characteristics, host plant species, and mycorrhizal fungi types, mycorrhizal fungi can be divided into arbuscular mycorrhizal (AM) and ectomycorrhiza (ECM). Ectoendomycorrhiza, orchid mycorrhiza, ericoid mycorrhiza, monotropoid mycorrhiza, and arbutoid mycorrhiza. Among them, ectomycorrhizal (ECM) and arbuscular mycorrhizal (AM) are the most diverse and widely distributed mycorrhizal types.

As early as 1885, the German botanist and plant physiologist B. Frank first discovered mycorrhiza, that is, the symbiosis of special fungi (mycorrhizal fungi) in the soil and plant roots [4], which opened a series of studies on the structure, type, formation mechanism, physiological and ecological functions of mycorrhizal fungi, and mycorrhizal biotechnology, forming an independent discipline-mycorrhizology [5, 6].

2. Diversity of mycorrhizal fungi

According to different host types and symbiotic characteristics, mycorrhizal fungi are divided into seven categories [5]: ectomycorrhizal (ectomycorrhizal; eCM), arbuscular mycorrhiza (arbuscular mycorrhiza; aM), ectoendomycorrhizas (ectoendomycorrhizas; eEM), orchid mycorrhiza (orchid mycorrhiza; oRM), arbutoid mycorrhiza (arbutoid mycorrhiza; aRM), ericoid mycorrhiza (ericoid mycorrhiza; eRM), and monotropoid mycorrhiza; mM) seven categories. According to the criteria of root tissue morphological differentiation and host plant lineage, mycorrhiza is divided into four main types: arbuscular mycorrhiza (arbuscular mycorrhiza; aM), ectomycorrhizal (ectomycorrhizal; eCM), ericoid mycorrhiza (ericoid mycorrhiza; eRM), and orchid mycorrhiza (orchid mycorrhiza; oRM). In mycorrhizal symbionts, mycorrhizal fungi absorb water and mineral elements in thc soil through mycelia and provide them to host plants. At the same time, host plants provide carbohydrates to mycorrhizal fungi in return. In addition, the functions of mycorrhiza in the ecosystem also include affecting the construction of aboveground plant communities, regulating soil carbon sequestration, nutrient cycling, ecosystem stability, and so on [7–11]. Mycorrhiza is widely distributed in the terrestrial environment. Under natural conditions, more than 80% (92% of the family level) of terrestrial plant roots can be infected by mycorrhizal fungi to form mycorrhiza [12]. Although most plants can form mycorrhizae, the distribution of mycorrhizae is not uniform. Ectomycorrhizae (2% of the total), arbuscular mycorrhizae (72%), and orchid mycorrhizae (10%) are the main mycorrhizae.

2.1 Diversity of arbuscular mycorrhizal fungi

Arbuscular mycorrhiza (AM) is the earliest and most widely distributed and is known as the mother of plant root symbionts. Therefore, it has attracted extensive attention from mycorrhizal scholars. Most (about 80%) terrestrial plant species can form this mycorrhiza [13]. Arbuscular mycorrhizal fungi (AM fungi), which can infect plant roots to form AM, are obligate nutritional fungi, and the carbohydrates required for their growth and development are all derived from host plants [6]. The structure of AM fungi is composed of two parts: inner root and outer root. The intraradical hyphae, vesicles, and arbuscules formed by AM fungi in the root cortex are located in the root cortex. The structures such as dendric or cauliflower-like branches belong to the root structure of AM fungi, while the extraradical hyphae and spores located in the root and soil together constitute its extra root structure.

2.1.1 Arbuscular mycorrhizal fungi

Arbuscular mycorrhizal (AM) is a symbiont formed by arbuscular mycorrhizal fungi (arbuscular mycorrhizas fungi, AMF) and plant roots. In addition to individual families such as Juncaceae, Caryophyllaceae, and Brassicaceae, which cannot or are not easy to form AM symbionts, most plants can be infected by AM fungi [6]. AM is also known as endomycorrhizas because it forms a symbiotic structure in the cortical tissue of plant roots. Its structural components include plant roots, soil hyphae outside the root, spores, auxiliary body, and intracellular and intercellular hyphae, vesicles, and arbuscules inside the root. AM was once known as vesicle-arbuscular mycorrhizas (VAM), but later studies found that only about 80% of AM fungi form vesicular structures, so the term VAM was discarded [14].

The species classification of AM fungi has always been an important aspect of AM fungal diversity research. By 2015, 254 AM fungal morphological species had been recorded worldwide (http://schuessler.userweb.mwn.de/amphylo/; statistics dated February 2015), which can be divided into 4 orders, 11 families, and 17 genera. Among them, there were 121 species of Glomeraceae, 40 species of Acaulosporaceae, 11 species of Diversisporaceae, 9 species of Ambisporaceae, 7 species of Pacisporaceae, 6 species of Claroideoglomeraceae, 5 species of Gigasporaceae, 4 species of Sacculosporaceae, 3 species of Paraglomeraceae, 2 species of Archaeosporaceae, and 1 species of Geosiphonaceae. At the same time, great progress has been made in the study of AM fungal species diversity in China. According to incomplete statistics, 113 species of AM fungi belonging to 7 genera have been isolated in China, including 22 species of Aeaulopora, 3 species of Entrophospora, 3 species of Archaeospora, 65 species of Glomus, 1 species of Paraglomus, 4 species of Gigaspora, and 15 species of Scutellospora [15].

Since 1974, the number of researchers on AM fungi has gradually increased. The corresponding researchers have described the morphological characteristics of some AM fungi and established the family Endogone. AM fungi are classified as Endogone [16]. In 1875, Berkeley and Broome [17] established the genus Sclerocystis. In 1889, the genus Sclerocystis and the genus Endocystis were placed under the family Endocystis established by Paoletti in 1889. In 1922, Bucholtz [18] indicated that the family Endogonaceae belongs to the order Mucormycota. In 1974, Gerdeman and Trappe [19] newly described Acaulospora and Megasporangium and reclassified Endomycaceae into seven genera: Glomus, Endogone, Modicella, Gigaspora, Acaulospora, Glaziella, and Sclerocystis. In 1979, Benjimin moved the family Enterobacteriaceae from Mucor to Endogonales [20]. The biggest change in the systematic classification of AM fungi was in 1999–2000, Morton and Benny [21] established Glomals, which consists of 2 suborders, 3 families, and 6 genera. Almeida and Schenck [22] transferred all the strains except *Sclerocystis coremioides* into Glomus. In 2000, Redecker [23] transferred them into Glomus and abolished Sclerocystis. Schublerr [24] analyzed the 18S r DNA sequence of AM fungi in 2002 and removed AM fungi from Zygomycota and reestablished Glomeromycot. The reason why the classification system of AM fungi changes so frequently is the limitation of morphological identification. The same species in different countries and regions have different life histories, and the results of morphological structure in different environments are quite different. It is this instability that promotes the direction of systematic classification gradually tends toward molecular methods. Oehl et al. [25] divided AM fungi into 1 class, 5 orders, 14 families, and 26 genera by combining modern molecular biology techniques and traditional morphological identification methods. So far, less than 250 species of AM fungi have been described, and only 341 species of AM fungi have been obtained by environmental DNA analysis. Such a small amount of AM fungi can form a symbiotic relationship with more than 200,000 plants on the earth, which means that AM fungi do not have strict host specificity

[26]. Therefore, the classification of AM fungi is still under development, and the classification of taxonomy is inconclusive.

AMF belongs to Glomeromycota in the classification system. According to the latest classification system of AMF [27], there are 1 class (Glomeromycetes), 4 orders (Glomerales, Diversisporales, Paraglomerales, Archaeosporales), 11 families (Glomeraceae, Claroideoglomeraceae, Gigasporaceae, Acaulosporaceae, Pacisporaceae, Diversisporaceae, Sacculosporaceae, Paraglomeraceae, Geosiphonaceae, Ambisporaceae, Archaeosporaceae), 29 genera (Dominikia, Funneliformis, Glomus, Kamienskia, Rhizophagus, Sclerocystis, Septoglomus, Claroideoglomus, Bulbospora, Cetraspora, Dentiscutata, Gigaspora, Intraornatospora, Paradentiscutata, Racocetra, Scutellospora, Acaulospora, Pacispora, Corymbiglomus, Diversispora, Otospora, Redeckera, Tricispora, Sacculospora, Paraglomus, Geosiphon, Ambispora, Archaeospora, Entrophospora), and about 300 species.

2.1.2 Inheritance of arbuscular mycorrhizal fungi

The sum of the genetic information carried by arbuscular mycorrhizal fungi is the genetic diversity of AM fungi. It was found that there were a large number of genetic variations among AM fungal populations and within populations, even within a single bosom. This shows that AM fungi have rich genetic diversity [28]. The generation of genetic diversity is based on changes in DNA genetic material, mainly reflected in gene polymorphism [29, 30]. Sanders et al. first reported the genetic information of Glomus fungi: DNA sequence polymorphism [31]; Lloyd-Macgilp et al. [32] studied and analyzed the ITS gene sequences of rDNA among five Glomus mosseae strains with different geographical origins, and found that there were sequence polymorphisms. Redecker et al. [23] studied the molecular system of AM fungi by PCR amplification of the 18S rRNA gene of G.spinuosum and Scleroystis coremioides and RFLP analysis of the products by restriction endonuclease digestion. Developmental status; Antoniolli et al. [33] amplified the ITS region of G.mosseae and Gi.margarita isolated from pasture after a single culture, and found that there were differences between the ITS regions of the two strains, and the degree of genetic variation of G.mosseae was greater than that of Gi.margarita. Pawlowska and Taylor [34] found that there were three kinds of ITS region variations in each nucleus of G.etunicatum. The genetic variation of AM fungi exists not only in ribosomal genes but also in the gene coding region, which is universal [34]. Kuhn et al. [35] found that the highly conserved binding protein gene (BiP) in fungi was mutated in the single offspring of G.intraradices; Corradi et al. [30] found that there were differences in the number of copies of the gene.

Changes in genetic materials can cause changes in AM fungal metabolism and differences in morphological characteristics. Bever and Morton [36] showed that nuclear migration and hyphal fusion occurred during the division of the mycelium of Sc.pellucida, resulting in nuclear recombination, which further led to a variety of differences in the single culture and the shape and size of the five single-saturated offspring were not the same. Koch et al. [37] reported that the AFLP amplification bands of G.intraradices populations were quite different under the same environmental conditions. These genetic variations have led to differences in physiological and symbiotic functions within AM fungal populations, such as differences in the length of AM fungal hyphae that further translate into differences in nutrient uptake by symbiotic plants [38].

At present, there are two hypotheses about the genetic variation of AM fungi. Kuhn et al. [35] reported that the genetic variation of AM fungi is likely to be due to the fusion of hyphae in the asexual reproduction of AM fungi, resulting in the formation of new hyphae and spores containing nuclei with different genetic components, and the frequency of intranuclear gene recombination is not high, that is, the genetic variation of AM fungi is distributed in different nuclei. However, Pawlowska and Taylor [34] believed that the abundant genetic variation in AM fungi was due to ploidy, that is, all the variations were contained in the mononuclear cells of AM fungi. Therefore, the genetic structure of AM fungi needs to be further studied. These rich genetic diversity should be considered in the study of physiological and ecological molecular biology of AM fungi, because they may be converted into functional diversity [28]. At the same time, the genetic diversity of AM fungi laid the foundation for the application of molecular biology analysis technology in the classification and identification of AM fungi and made the classification of AM fungi more scientific and perfect.

The diversity of AM fungi is the result of the combined action of various factors. Host plant types, plant compounds, and environmental factors can affect the diversity of AM fungi. Tawaraya et al. [39] confirmed that the host plant type of AM fungi is one of the important factors affecting the distribution of AM fungal species. The composition and infection rate of AM fungal community vary with different host plants. Some scholars believe that the influence of plant species on the diversity of AM fungal community is mainly due to the effect of plant root metabolites. Alkaloids and flavonoid metabolites synthesized by plants will affect the germination of AM fungal spores. The research directions of AM fungal diversity at home and abroad have different emphases. Most of the articles focus on soil physical and chemical properties and AM fungal diversity. Zhang Haibo [40] RDA analysis showed that plant species had a certain effect on the difference of AM fungal community structure. The composition of AM fungal community in different plant rhizospheres was different, indicating that AM fungi had specific selection for host plants. Yang Xiuli et al. [41] found that the distribution of AM fungal species richness in different seasons in the same forest type was regular, and the highest species richness was in August. The distribution of AM fungal spore density in different seasons also showed a certain regularity, and the highest spore density was in September. Combined with the research on the diversity of AM fungi in different forest types, it was found that the composition of AM fungi in different forest types was different, the spore density was different, and the dominant species were different. AM fungi and forests have undergone a long period of evolution, and the entire ecosystem has now entered a complete and orderly positive feedback virtuous cycle. The use of AM fungi to promote plant colonization growth and vegetation restoration should consider its species preference. Therefore, studying the relationship between AM fungi and forest types is helpful in protecting AM fungi resources and forest resources.

2.1.3 Origin of arbuscular mycorrhizal fungi

Based on existing fossil evidence, AM symbionts originated in the Ordovician (Ordovician) at least 460 million years ago and play an important role in the early plant login process and subsequent evolution of terrestrial plants [12, 42, 43]. Plants face many challenges in the early stage of landing, such as arid terrestrial conditions, nutrient-poor soil, and strong ultraviolet radiation. With the help of AM fungi, plants can effectively absorb soil mineral nutrients and have the ability to resist adverse environmental conditions [6]. Therefore, the emergence of AM fungi is considered to be an important prerequisite for the successful early landing of plants [43]. Although this assertion can be corroborated by recent research, the study found that the earliest origin of a class of plant land money (liverwort; it originated about 430 million years ago) and can be infected by AM fungi, which significantly enhances the nitrogen and phosphorus absorption capacity of *Marchantia polymorpha*, and increases its growth rate and fitness [44]. However, how was the early plant-AM fungal mutualism established? This remains to be further explored and scientifically answered.

AM fungi is an ancient asexual reproduction of the biological groups [45], in its long-term evolution process, multinuclear genome slowly emerged, that is, there are multiple nuclei appear at the same time in its hyphae and asexual spores, a single spore can appear up to tens of thousands of nuclei [35, 46]. In the process of reproduction of AM fungi, many nuclei can be transferred to the mycelium or spores of the offspring, so that AM fungi have extremely high individual genetic diversity, and often have high genetic diversity in terms of individual spores [31, 34, 47]. It is precisely because of its asexual reproduction and multinuclear characteristics, although the species differentiation rate of AM fungi is slow, it also suggests that it may have a faster functional differentiation rate, and the intraspecific functional diversity of AM fungi should be high [48–50]. At the same time, the existence of multinuclear genomes is bound to bring greater resistance to the molecular identification of AM fungal species that are currently widely used [31]. High genetic variability to some extent hinders us from identifying AM fungal diversity through DNA sequences. Although the characteristics of asexual reproduction of AM fungi have been widely recognized, new research evidence shows that a large number of specific genes related to sexual reproduction are present in AM fungi [51, 52]. Although these sexual reproduction-related genes cannot prove that sexual reproduction exists in AM fungi, they provide new ideas and directions for future research, which will help us fully understand AM fungi and understand their genetic and evolutionary processes [53]. In the recently completed genome sequencing of the representative species of AM fungi (Rhizophagus irregularis), 28,232 genes were found [54]. Further annotation and analysis of these genes open the door for future research on the genetics, evolution, symbiotic mechanism, and physiological function of AM fungi.

2.1.4 Influencing factors of arbuscular mycorrhizal fungal diversity

Because of the mutual selection relationship between AM fungi and host plants, and the strong adaptability of AM fungi to different ecological environments, the diversity of AM fungi shows obvious differences in different ecosystems [26, 55, 56] and is affected by many factors at the same time.

2.1.4.1 Host plants

Because AM fungi rely on obtaining nutrients from host plants to maintain their own survival and reproduction, the structure and diversity of AM fungal communities are bound to be affected by host plant communities [57]. A large number of studies have shown that the community structure of host plants will affect the structure, function, and diversity of AM fungal community in root-soil to a certain extent [58, 59]. Hausmann et al. found that the nutritional needs of host plants at different stages of growth will affect the AM fungal community structure in their roots [60].

2.1.4.2 Soil factors

AM fungi exist in the root tissue and rhizosphere soil of host plants, so the salinity, pH, and nutrient elements of soil will affect the structure and diversity of AM fungal community. At the same time, AM fungi will also change the physical and chemical factors of rhizosphere soil [60]. Soil nutrients can affect the colonization, growth, distribution, and spore production of AM fungi. Among them, phosphorus is the most closely related to AM fungi in all soil nutrients. Studies have shown that when the content of available phosphorus in rhizosphere soil is low, that is, under the condition of phosphorus as a limiting condition, the addition of phosphorus fertilizer will appropriately promote the reproduction of AM fungi. However, when enough phosphorus fertilizer is added and the phosphorus limiting conditions of the soil are lifted, the competition of the aboveground part of the plant will be intensified, resulting in a decrease in the carbon content transported by the plant to the underground roots and mycorrhiza, thus inhibiting the spore production and colonization of AM fungi. In turn, it affects the structure and diversity of AM fungal communities [61, 62]. Similar to phosphorus, soil nitrogen also has an important effect on the community structure and diversity of AM fungi [63]. The simultaneous changes of nitrogen and phosphorus will change the distribution of carbon content in the aboveground and underground parts of the plant, which will reduce the diversity of AM fungal communities [64, 65].

2.1.4.3 Climatic factors

Climatic factors mainly affect the growth and reproduction of AM fungal community through temperature, light, and water conditions. Studies have shown that soil temperature can directly affect the ability of AM fungal community to infect host plants. When the ambient temperature is low, AM fungi will produce more vesicle structure. However, when the temperature increases, the colonization rate and mycelial structure of AM fungi increase significantly [66].

2.1.4.4 Geographical factors

There are significant differences in the structure and diversity of AM fungal communities in different environmental soils, mainly due to changes in altitude, elevation, and latitude. Geographical factors mainly affect AM fungi by indirectly affecting other environmental factors such as light, precipitation, and temperature. Cai Xiaobu et al. [67] found that the diversity and spore density of AM fungi increased with the increase of altitude in the Qinghai-Tibet Plateau at an altitude of 3700–5220 m.

2.2 Diversity of ectomycorrhizal fungi

ECM is an ectomycorrhizal fungus that forms mycorrhizal symbionts with the roots of host plants. The hyphae of ECM fungi will wrap the surface of the host root tip to form a mantle, but their hyphae do not penetrate the interior of the plant cells, but only extend and grow between the cells to form a Hartig net. Emanating hyphae are sparse or dense, with special mycorrhizal symbiotic structures such as fungal mantle, Hartig net, and epitaxial hyphae, which are called ectomycorrhizal fungi [68, 69]. These symbiotic fungi are called ectomycorrhizal fungi (EcMF).

This symbiotic relationship of ECM fungi in plant roots has been widely studied. According to statistics, there are more than 20,000 fungi in the world, which can form ECM symbionts with 6200 plants [70]. This symbiotic relationship has a wide impact on forest ecosystems. Fagaceae [71] (Fagaceae), Pinaceae [72] (Pinaceae), Betulaceae [73] (Betulaceae), Myrtaceae [74] (Myrtaceae), Juglandaceae [75] (Juglandaceae), Salicaceae [76] (Salicaceae), and other dominant tree species in forest ecosystems have important economic and ecological values. The symbiotic relationship between these plants and ECM fungi is one of the key factors for their normal growth, production, and reproduction [77].

The formation of ectomycorrhizal (ECM) usually includes four stages: contact (recognition), invasion, expansion, and emergence. These stages are usually completed in a short time, and there is no obvious boundary between them [78]. In general, the formation of ECM is initiated by the proliferation of fungi on the root surface caused by the root exudates of the host plant. The mechanism of fungal invasion into plant tissues involves a variety of complex biological processes, including the stacking and interweaving of hyphae, as well as the expansion and enclosure of hyphal networks [79]. After the formation of the mantle, the fungi continue to expand inward, through the root epidermis into the cortex tissue and grow rapidly, forming a Hastelloy network, which marks the completion of the invasion stage. In this process, fungi inhibit the defense response of host plants by secreting effector proteins [80]. It was found that there was signal communication in the symbiosis process between plants and ECM fungi. Plants transmit signals by secreting flavonoids [81, 82], while ECM fungi transmit signals by secreting auxin [83, 84]. Auxin can enter root cells, induce plant auxin signaling pathway, inhibit root growth, and induce lateral root formation. In addition, symbiotic fungi can release sesquiterpenes to promote plant lateral root formation, which is independent of the host plant auxin signaling pathway. Therefore, the mechanism of ECM formation is very complex, involving the interaction of multiple biological processes and signaling pathways [85].

Researchers have conducted extensive research on ECM, mainly involving the morphology of ECM, the promoting effect of ECM fungi on the growth of host plants, the diversity of ECM fungal fruiting bodies, and taxonomy. At present, it is known that about 280 genera [86–88] more than 20,000 species [77] of fungi can form ECM with 250–300 genera and 6000–7000 species of plants [89], mainly distributed in Ascomycota (Turberaceae, Gloniaceae, Helvellaceae, and Pezizaceae). Russulaceae, Cantharellaceae, Thelephoraceae, Atheliaceae, Pisolithaceae, Sclerodermataceae, Cortinariaceae, Clavariaceae, Boletaceae, Inocybaceae, Sebacinaceae, Amanitaceae, and Hymenogastraceae of Basidiomycota [90].

2.2.1 Diversity of ectomycorrhizal fungi in China

Chinese scholars have made great progress in the investigation of ectomycorrhizal fungi resources. The species, geographical distribution, and ecological characteristics of ectomycorrhizal fungi in the main afforestation tree species and important timber tree species in various forest areas have been basically clarified [91]. According to incomplete statistics, nearly 600 species of ectomycorrhizal fungi have been reported in China, belonging to 2 subphyla, Ascomycotina and Basidiomycotina, 28 families, and 63 genera. Chinese scholars have carried out many investigations on pine species, mainly including *Pinus yunnanensis*, *Pinus massoniana, Pinus densiflora, Pinus sylvestris var.mongolia, Pinus taiwanensis*, and *Pinus koraiensis*. More than 184 species of ectomycorrhizal fungi have been reported. It belongs to 20 families and 41 genera. Among them, Russula, Inocybe, Tomentella, Cortinarius, Amanita, Boletus, and Pisolithus are common fungi. In addition, some fungi also form ectomycorrhiza with coniferous forest types, such as larch (Larix) and spruce (Picea) and some broad-leaved forest types, such as Quercus, Eucalyptus, and Populus. Wang et al. [92] used the method of field ecology to investigate the ectomycorrhizal fungi of important afforestation tree species such as *Pinus tabulaeformis, Pinus massoniana*, and *Pinus sylvestris* in Northeast China, Central China, South China, and East China. A total of 19 species of ectomycorrhizal fungi were found.

In Northeast China, Zhao et al. [93] identified 51 species of ectomycorrhizal fungi of Russulaceae in the Changbai Mountain area through specimen collection and investigation. Meng Fengrong and Shao Jingwen [91] investigated the ectomycorrhizal fungi of 12 main coniferous forest types in the main forest areas of Changbai Mountains, Xiaoxing 'an Mountains, and Daxing 'an Mountains by means of field investigation and specimen collection and identification. There were 163 species of ectomycorrhizal fungi belonging to 43 genera and 19 families. Wang Shuqing and Xu Lihua [94] classified and identified 76 species of ectomycorrhizal fungi associated with larch, *Pinus sylvestris var.mongolica, Pinus koraiensis*, *Pinus tabulaeformis*, poplar, and other major timber species in Northeast China through field investigation and indoor microscopic examination. Tu Liguer et al. [95] obtained 100 species of ectomycorrhizal fungi in five typical vegetation communities of broad-leaved Korean pine mixed forest in Changbai Mountain by sampling survey method and morphological classification method. Wang [96] found that there were 16 species of underground ectomycorrhizal fungi of *Quercus mongolica* in Bingla Mountain of Liaoning Province by combining morphological and molecular biological methods. Song Fuqiang et al. [97] preliminarily identified 21 species of ectomycorrhizal fungi associated with *Pinus koraiensis* in Xiaoxing 'an Mountains, belonging to 5 families and 8 genera. Lang Qinglong et al. [98] investigated the ectomycorrhizal fungal resources of oak trees in Fengcheng area of Liaoning Province after more than 100 field investigations and specimen identification, and initially identified 35 species in this area. Wang Hui et al. [99] determined that there were 36 species of ectomycorrhizal fungi belonging to 7 families and 13 genera in Dandong area of Liaoning Province by fruiting body investigation and specimen identification. Among them, 29 species of ectomycorrhizal fungi were associated with *Quercus mongolica*. Cui and Mu [100] studied the resources of ectomycorrhizal fungi in the typical distribution area of Tilia amurensis in Heilongjiang Province by morphological and anatomical characteristics, and preliminarily divided them into 18 types.

In Northwest China, Tang et al. used the investigation and identification of fruiting bodies, observation of external morphology and anatomical structure to investigate and study 35 species of poplar in Shaanxi Province, and identified 9 species of ectomycorrhizal fungi [101]. Tian Maolin [102] investigated and identified 63 species of Russulaceae ectomycorrhizal fungi in Gansu Province. Wu et al. reported 78 species of ectomycorrhizal fungi in Shaanxi Province through field investigation and specimen collection [103]. Wu Chonghua et al. found that 28 species of ectomycorrhizal fungi were obtained by investigating the ectomycorrhizal fungi of the main tree species such as Abies fargesii and Larix chinensis in Taibai Mountain Nature Reserve [104]. Pu Xun reported that there were 77 species of ectomycorrhizal fungi symbiotic with trees in Longnan area of Gansu Province, belonging to 14 families and 28 genera [105]. Xun [106] used plant taxonomy and ecological methods to identify 37 species of ectomycorrhizal fungi in Helan Mountain, Ningxia [106]. Zhang Ruqin et al. [107] studied the ectomycorrhizal fungi under pine-oak mixed forest, larch forest, *Pinus tabulaeformis* forest, and *Pinus armandii* forest in Huoditang forest area of Qinling Mountains by field investigation, specimen collection, isolation, and identification. A total of 29 species, 16 genera, and 10 families [107] were identified. Qi et al. reported 28 species of ectomycorrhizal fungi in the middle part of Qinling Mountains (Shaanxi section of Qinling Mountains) by means of field investigation, specimen collection, culture, and isolation [108]. Ding et al. found that there were seven species of ectomycorrhizal fungi in Zibai Mountain Nature Reserve of Qinling Mountains by collecting specimens, morphological structure, and microscopic observation [109]. Through field investigation, specimen collection, and molecular biology, Guli Ahmati et al. [110] conducted a comprehensive investigation and study on ectomycorrhizal fungi in the Habahe plain area of Altay, Xinjiang, and identified 14 genera (21 species) of ectomycorrhizal fungi [110].

In Central China, Chen Zuohong et al. investigated the ectomycorrhizal fungi of Amanita in Mangshan Nature Reserve of Hunan Province by special investigation and collecting a large number of specimens. A total of 25 species have been identified [111]; Li Jianzong discovered two new records of ectomycorrhizal fungi (Amanita) in the mixed forest of *Pinus massoniana* and broad-leaved trees in Changsha, Hunan Province, and made specimens [112]; Cui et al. collected specimens and identified 31 species of ectomycorrhizal fungi of Russula in Henan mountainous area [113]. Zhang Linping et al. found 28 species of ectomycorrhizal fungi in Jiangxi Tongboshan Nature Reserve by macroscopic characteristics, microscopic observation, and specimen identification [114]. Li Jianzong et al. identified 25 species of ectomycorrhizal fungi in Shunhuangshan Nature Reserve by investigating and collecting specimens [115]. Li Jianzong et al. [116] found that there are 39 species of fungi that form ectomycorrhizal fungi with trees in Liubuxi Nature Reserve of Hunan Province by investigating and collecting specimens. Huang et al. [117] used morphological and molecular biological identification methods to investigate the ectomycorrhizal fungi of *Pinus massoniana* in lead-zinc mining areas in Central China and found that there were 47 species of ectomycorrhizal fungi symbiotic with *Pinus massoniana* [117].

In North China, Qin et al. [118] investigated *Castanea mollissima* in Huairou County, Yanshan Mountains, Beijing by collecting fruiting bodies, and identified 13 genera and 29 species of ectomycorrhizal fungi symbiotic with *Castanea mollissima* [118]. Cai Huaifu et al. collected specimens and indoor identification, and reported 7 species of ectomycorrhizal fungi collected in some areas of Beijing Songshan Nature Reserve [119]; Tian et al. identified 38 species of ectomycorrhizal fungi of Russulaceae in Chifeng, Inner Mongolia [120]. Fan Yongjun et al. used morphological anatomy and molecular biology methods to study the ectomycorrhizal fungi of *Betula platyphylla* in Inner Mongolia, and identified 13 species of ectomycorrhizal fungi associated with it [121]; Bai Shulan et al. identified 78 species (15 families, 31 genera) of ectomycorrhizal fungi in Daqing Mountain and 79 species (15 families, 32 genera) in Manhan Mountain by outdoor investigation and indoor observation [122]; Fan et al. studied and investigated Qinghai spruce in Helan Mountain area of Inner Mongolia from the perspective of morphology, anatomy, and molecular biology, and identified 11 different types of ectomycorrhizal fungi [123]. Bai et al. collected fruiting bodies for morphological and anatomical observation and initially believed that there were 163 species of ectomycorrhizal fungi distributed in Daqing Mountain, Inner Mongolia [124].

In South China, Gong Mingqin and Chen Yu [125] obtained 11 species of ectomycorrhizal fungi from Pinus and Anacardium in South China by combining outdoor investigation with indoor observation [125]. Rao Benqiang studied the ectomycorrhizal fungi of *Pinus massoniana* in Wuyishan Nature Reserve by macroscopic, microscopic, and ultrastructural methods. A total of 78 species of ectomycorrhizal fungi associated with *Pinus massoniana* were observed [126]. Qian et al. collected a total of 123 species of ectomycorrhizal fungi (84 species were identified) after 6 years of investigation on the plant community of Tsuga chinensis in Wuyishan Nature Reserve through macroscopic morphological observation, microscopic and ultrastructural observation [127].

In East China, Li Haibo et al. investigated the resources of ectomycorrhizal fungi in Lishui, Zhejiang Province through field investigation, collection, isolation, and culture, and identified 38 species of ectomycorrhizal fungi [128]. Ke Lixia and Liu Birong investigated the resources of ectomycorrhizal fungi under pine forests in Huangshan area by means of field ecological investigation, indoor observation, and microscopic examination, and identified 43 species of ectomycorrhizal fungi [129]. Chen Tianxing and Chen Shuanglin obtained 49 species of ectomycorrhizal fungi in Purple Mountain of Nanjing by collecting specimens and morphological taxonomy [130].

In Southwest China, Bi Guochang et al. investigated the ectomycorrhizal fungi in alpine coniferous forests in Northwest Yunnan by field investigation and collection of fruiting body specimens and identified more than 140 species of ectomycorrhizal fungi belonging to 3 genera [131]. He Shaochang has collected and identified 138 species of ectomycorrhizal fungi in Guizhou trees [132]. Tan Fanghe and Wang Yunzhang [133] investigated the resources of ectomycorrhizal fungi under pine forests in Sichuan Province by investigating fungal fruiting bodies, collecting specimens, indoor observation, and microscopic examination, and identified 9 families, 18 genera, and 50 species [133]. Yu Fuqiang et al. [134] collected fruiting body specimens and observed microscopic characteristics, and investigated the ectomycorrhizal fungi under the main Pinus yunnanensis forests in Yunnan Province. In total, 211 species of ectomycorrhizal fungi belonging to 39 genera and 27 families were identified.

Zhu Tianhui et al. conducted a systematic investigation and study on the species of ectomycorrhizal fungi of Eucalyptus in Sichuan Province through field investigation and indoor identification. It was found that there were 17 types of ectomycorrhizal fungi associated with Eucalyptus, belonging to 9 families and 11 genera [135]; Liu Xiaojiao et al. collected specimens and consulted relevant literature, and sorted out 23 species of ectomycorrhizal fungi of Russulaceae in Sejila Mountain, Tibet [136]; Zhou Guanglin et al. collected specimens and indoor identification, and found that there are 37 species of ectomycorrhizal fungi in Chongqing Jinyun Mountain National Nature Reserve [137]; Luo et al. found that there were 68 species of ectomycorrhizal fungi in Pinus massoniana in Kaili City, Guizhou Province through field investigation [138].

2.2.2 Taxonomy of ectomycorrhizal fungi

At first, scientists mainly identified the species of ECM fungi by morphology and studied their diversity. In recent years, molecular biology techniques have been widely used in the study of ectomycorrhizal fungi, and more and more ECM fungi have been identified. According to the latest statistics, there are about 20,000–25,000 species of ECM fungi worldwide, mainly from Basidiomycota and Ascomycota, and a small amount from Zygomycota [139].

According to the life form and phylogenetic characteristics of ECM fungi, scientists divided all ECM fungi into 66 evolutionary branches (lineages), including 37 evolutionary branches of Basidiomycetes, 27 evolutionary branches of Ascomycota, and 2 evolutionary branches of Zygomycota [140]. Wurzburger et al. identified that

Suillus tomentosus, Cenococcum geophilum, and an ECM fungus from Russula were widely distributed in the swamp forest transition zone in southeastern Alaska [141]. Smith et al. studied the ECM fungal community of three leguminous tree species in tropical rainforests. A total of 118 ECM fungi were identified and divided into four branches [142]. Lamit et al. identified 367 ECM fungi in three different elevation gradients in the forests of Northern Iran, and the abundance of ectomycorrhizal fungi decreased with increasing altitude [143]. Bonito et al. identified 44 genera of ECM fungi in *Carya illinoinensis* (*Carya illinoinensis*) [144].

In recent years, Chinese scholars have also applied molecular biotechnology to the study of ECM fungal diversity. The dominant genera of ECM fungi in *Pinus koraiensis* pure forest in Northeast China were Sebacina, Suillus, Meliniomyces, Russula, Tomentella, Rhizopogon, and Amphinema [145]. The ECM fungi of natural *Pinus sylvestris var.mongolica* of different age classes in Honghuaerji belong to 2 phyla, 4 classes, 12 orders, 26 families, and 43 genera. The proportion of dominant genera of ECM fungi in different age classes is different [146]; the ECM fungi in the rhizosphere soil of *Betula platyphylla* in Inner Mongolia came from 2 phyla, 6 classes, 11 orders, 25 families, and 38 genera [147]; ECM fungi in 15 Pinus tabulaeformis forests in 9 provinces of Northern China belong to 2 phyla, 5 classes, 13 orders, 24 families, and 41 genera [148]; the ECM fungi of the main oaks in the Qinling Mountains and the Loess Plateau belong to 2 phyla, 6 classes, 13 orders, 28 families, and 48 genera [149]; the ECM fungi associated with Betula albo-sinensis in the Qinling Mountains belong to 6 orders, 8 families, and 9 genera, and the ECM fungi associated with *Larix gmelinii* come from 4 orders, 11 families, and 11 genera [150]; eCM fungi belong to 3 phyla, 8 classes, 18 orders, 23 families, and 18 genera in Fagaceae of tropical mountain rainforest [71].

2.2.3 Investigation of ectomycorrhizal fungi resources

2.2.3.1 Investigation of ectomycorrhizal fungi resources abroad

For more than a century, mycorrhizal researchers at home and abroad have done a lot of work in the investigation of ECM resources, and more and more ECM fungal resources have been detected and identified. The diversity of ECM fungi has been investigated abroad for a long time. As early as 1889, Noack tried to detect the resources of mycorrhizal fungi by tracing the mycelium of fruiting bodies [151]. Subsequently, Melin [152, 153] preliminarily speculated that there was a surprising number of ECM fungal diversity under natural conditions by investigating the occurrence of fruiting bodies in the field [152]. Since the 1990s, mycorrhizal research has become a hot spot. Systematic and large-scale investigations have confirmed the speculation of Melin et al. mycorrhizal researchers also continue to predict the global ECM fungal diversity based on the types of ECM that have been found. First, according to statistical analysis by Molina et al. [154], he predicted that there are about 5000–6000 ECM fungi in the world [153]. As many fungal species have been shown to form ECM with plants, Taylor and Alexander [155] reestimated the diversity of ECM fungi and believed that there are about 7000–10,000 ECM fungi in the world [154]. In recent years, with the use of molecular biology methods for ECM diversity research, this number has further expanded. The latest statistics show that about 20,000–25,000 fungi can form ECM with about 6000 higher plants around the world [139, 155]. The current research results show that ECM fungi are derived from three fungal phyla, namely Basidiomycota, Ascomycota, and Zygomycota [156, 157].

The proportion of Basidiomycota is the largest, reaching about 95%, followed by Ascomycota, about 4.8%, and only a small part of the fungi of Zygomycota can form ECM [154]. According to the life form and phylogenetic characteristics of ECM fungi, Tedersoo et al. divided all ECM fungi into 66 evolutionary branches (lineages), including 37 evolutionary branches of Basidiomycetes, 27 evolutionary branches of Ascomycota, and 2 evolutionary branches of Zygomycota [87].

2.2.3.2 Investigation of ectomycorrhizal fungi resources at home and abroad

The study of ECM fungi in China started late but developed rapidly. In the early stage, the investigation of ECM resources in China was mainly based on one or more tree species on the local scale. Bi Guochang et al. conducted a systematic investigation of ECM fungal resources in alpine coniferous forests in Northwestern Yunnan using a combination of standard plot surveys and route surveys. A total of 33 genera and 140 species of ECM fungi were identified [131]; Chen Lianqing found that a total of 27 fungi can form ectomycorrhizal fungi with *Pinus massoniana* [158]; Gong Mingqin and Chen Yu investigated ECM fungi from two hosts in South China, and found 11 ECM fungi [125]. Zhao Zhong et al. [159] investigated the ECM of *Populus tomentosa* in the central plains and found a total of 14 ECM fungi [159]. Tang et al. [101] investigated 35 ECM fungal resources of poplar in Shaanxi Province and identified 9 ECM fungi. Wu et al. [104] investigated the ectomycorrhizal resources in the Taibai Mountain Nature Reserve and identified 28 species of ectomycorrhizal fungi, belonging to 6 families and 18 genera [104]. Meng and Shao investigated the underground ECM of the main coniferous forests in Changbai Mountain, Lesser Khingan Mountains, and Greater Khingan Mountains in Northeast China, and finally identified 19 families, 43 genera, and 163 species of ECM fungi by means of indoor identification [91]. Bai Shulan conducted a systematic investigation of ECM fungal resources in Daqingshan area of Inner Mongolia. A total of 163 ECM fungi were found, belonging to 18 families and 41 genera [160]; Fan Yongjun analyzed ECM fungal resources of four host plants in Inner Mongolia from the perspective of morphology and anatomy and finally identified 48 mycorrhizal morphological types [161].

Later, with the development of molecular biology techniques, Chinese mycorrhizal researchers kept pace with the times and quickly used this technique to study ECM diversity, and the spatial scale of the study also expanded from the early local scale to the regional scale. Zhang found that 19 kinds of fungi can form ECM with *Pinus sylvestris var.mongolica* [162]. Wang Qin et al. studied the diversity of ECM fungi in temperate Quercus liaotungensis and subtropical Castanopsis fargesii, and detected 66 and 17 OTUs from each of the two host plants [163, 164]. Ding et al. found 36 ECM fungi in the subtropical evergreen broad-leaved forest in Southwest China [165]; Yi et al. used high-throughput sequencing to detect a total of 143 OTUs in Qinghai spruce forest in Helan Mountain, Inner Mongolia, belonging to 20 families and 25 genera [166]. Geng Rong et al. [167, 168] investigated the ECM fungi of Picea asperata and Quercus aliena var.acuteserrata in Xinjiashan forest area of Qinling Mountains by combining morphological and ITS sequencing methods. The results showed that 37 and 51 species of ECM fungi were detected, belonging to 10 families and 14 genera, respectively [167, 168]. Gao et al. used 454 pyrosequencing to find 393 OTUs in different succession stages of forests in Gutian Mountain, Zhejiang Province, belonging to 21 ECM fungal evolutionary lineages [169]. Wen Zhugui et al. conducted a systematic study on the ECM fungi of Pseudotsuga sinensis, an endangered species in China.

A total of 66 fungi were identified to form ECM with Pseudotsuga sinensis [170]; Han Qisheng et al. detected a total of 60 ECM fungal OTUs in the alpine timberline species of *Larix gmelinii* forest in Taibai Mountain, belonging to 16 families and 28 genera [171]. Wang et al. combined morphological and molecular identification to carry out large-scale sampling and research on ECM fungal communities in five different habitat types of Quercus liaotungensis forests in Northern China. A total of 220 OTUs were detected to belong to 28 ECM fungal evolutionary branches [172].

3. Mycorrhizal fungal function

3.1 The ecological function of AM fungi

The ecological functions of AM fungi are reflected in the stability of plant communities and plant ecosystems and the improvement of ecosystem productivity [173]. The ecological functions of AM fungi are mainly applied to the restoration of degraded soil and the reconstruction of degraded vegetation [174]. Studies have shown that the inoculation of AM fungi significantly improved the ability of host plants to repair heavy metal-contaminated soil. AM fungi infect host plants to establish a symbiotic system, enhance plant tolerance to heavy metals, and promote plant growth and metabolism by chelating soil heavy metal ions with extracellular hyphae and expanding root absorption area [175]. It affects the morphology of heavy metals in soil and regulates the transport of heavy metals in plants by affecting the secretion effect of roots and changing the absorption kinetics of roots [176]. Wang Haijuan et al. used alfalfa to inoculate AM fungi. After inoculation, it was found that it could significantly improve the root growth of alfalfa and improve the rhizosphere soil microorganisms [177]; Zhang Zhongfeng et al. reported that inoculation of AM fungi in karst areas could promote the growth and biomass growth of Cyclobalanopsis glauca seedlings. The above studies have found that the inoculation of AM fungi has a certain potential effect on the stability of vegetation in the ecosystem and the recovery ability of the natural ecosystem [178].

3.2 The role of mycorrhizal fungi in terrestrial ecosystems

3.2.1 Symbiosis with plants

Mycorrhizal fungi can form symbionts with plants. Mycorrhizal fungi in addition to part of the fungal hyphae in the root surface of the formation of hyphal sheath or into the root cortex cells to form vesicles from the branches, there are a large number of hyphae stretching in the rhizosphere soil of plants, these stretching to the rhizosphere soil of the mycelium to expand the absorption area of plant roots, enhance the absorption capacity of mineral elements and water in plants, through isotope tracing and water determination, has proved that mycorrhizal fungi can promote the absorption and utilization of P, N, K, Cu, Zn, and other elements and water in plants [179–181]; on the other hand, fungi also obtain carbohydrates and other nutrients through the roots of plants, thus forming a mutually beneficial relationship in nutrition.

3.2.2 Entering the food web as food

Many fungi that form ectomycorrhizal fungi can form fruiting bodies with different shapes and large volumes on the surface or trunk in the wet rainy season, and some can form truffles in the ground. Many of these fruiting bodies or truffles are the food of arthropods, some small mammals and some birds in the soil ecosystem, and many of them are delicious foods for humans, such as fungi of the genera Cantharellus, Boletus, Lactarius, Tricholoma, Russula, and Ramaria. On the one hand, these fungi play a role in maintaining species diversity in the ecosystem as a food source for animals; on the other hand, the biomass of mycorrhizal fungi directly enters the food web through animals, thereby accelerating the circulation of substances in the ecosystem.

3.2.3 Affecting the succession of plant communities and the floristic composition of communities in terrestrial ecosystems

In the fifties and sixties of the last century, ecologists have noticed that in the process of natural restoration of man-made or natural damaged ecosystems, the plants that first settled in the damaged habitats are often not the dominant species before the destruction; the more serious the damage to the ecosystem, the slower the natural recovery. In 1965 Mayr argued: "Why can't settle successfully in unspoiled habitats where there is usually strong competition for settlers in damaged habitats?" Why is it particularly easy to settle in disturbed habitats? May's own answer is that " the adaptive advantages of these species under certain conditions are usually at the expense of losing the adaptability of some other adaptive components and types," which is obviously a self-contradictory argument. After that, ecologists studied the flora composition and succession of newly formed volcanic islands, such as Long Island and Iceland near Guyana. It was found that most of the plants that first settled on the volcanic island were Polygonaceae, Amaranthaceae, Cruciferae, Urticaceae, and Caryophyllaceae. At any time, plants of other families gradually moved to settle [182]. These first settled families are recognized as families that do not form mycorrhiza under natural conditions [19, 183]. In 1979, Reeves et al. published their comparative study of mid-elevation sage communities in natural and damaged ecosystems and found that 99% of plants in natural communities had mycorrhiza [184], while in the damaged ecosystem adjacent to the natural community (road embankment), the number of plants with mycorrhiza was less than 1%. According to the results of this study, and with reference to the settlement and succession of plants on the volcanic island, Reeves et al. made a reasonable explanation for the problem raised by Mayr. They believed that in the damaged ecosystem, the disturbance to the soil may reduce or eliminate the propagules of mycorrhizal fungi (generally referring to surviving hyphae and spores); the reduction of mycorrhiza fungal propagules reduced the possibility of infection of new host plants, and non-mycorrhizal species successfully colonized and established communities (because plant species that require mycorrhizal nutrition died at the seedling stage due to colonization of mycorrhiza fungal propagules on their roots even after immigration). The non-mycorrhizal species further grew successfully, and further reduced the propagules of mycorrhizal fungi (the survival time of mycorrhizal fungi in the absence of host plants was limited). All mycorrhizal fungal propagules were eliminated, and competition with mycorrhizal plant species was also avoided. It is very slow to return to the system state before being disturbed (only by the slow invasion of mycorrhizal fungi). The more serious the interference, the greater the possibility of eliminating mycorrhizal fungal propagules, so the recovery of the system will take longer. This explanation of Reeves et al. was further confirmed by later studies. Janos 's study showed that in the humid tropics, the restoration of disturbed ecosystems was first settled by facultative mycorrhizal plants in the damaged environment, and then gradually replaced by mycorrhizal plants and developed into a climax ecosystem. These results indicate that mycorrhizal fungi play an important role in the succession type, succession speed, and flora composition of plant communities in terrestrial ecosystems.

3.2.4 Regulation of resource allocation in ecosystems

In natural terrestrial ecosystems, the vast majority of plants form mycorrhizas, and the fungi that form mycorrhizas with plants are usually less specific to their hosts [185]. This is especially true for mycorrhizal fungi from branches [186]. Therefore, in the undisturbed forest ecosystem, the hyphae that grow outward from the roots of infected plants can further infect the roots of other plants, so that some unrelated plants in the system can be linked together through the hyphae of different mycorrhizal fungi, and the material and information can be exchanged through the hyphal network or hyphal bridge or hyphal connection between plant roots. Ecologists have noticed this way of binding in ecosystems and referred to these plants bound together by hyphae as "Guilds." It has been proved that plants in dependent plant groups can use hyphae as channels to transmit C, N, P, and other substances between plants in two directions [185], thus affecting the allocation of resources in the ecosystem. Grime et al. further studied the role of this hyphal network in maintaining species diversity in ecosystems, and through research under experimental conditions, it is believed that this hyphal network shared by different plant species can increase species diversity in ecosystems by bidirectionally transferring carbohydrates [187]. However, Bergelson et al. considered that the study under this experimental condition did not prove that there was a relationship between "source" and "sink" of carbohydrates between the two plants, nor did it indicate that there was a "net flow" of carbohydrates between plants [188]. Therefore, this study could not be of universal significance.

3.2.5 Promote the absorption and utilization of water and nutrients by host plants

ECM mainly promotes the absorption of nutrients and water by expanding the absorption area and absorption range of tree roots. A large number of studies have shown that under forest site conditions, plant mycorrhizal fungi have different types of hyphal morphology in the soil. These hyphae can form a large hyphal system. The hyphal systems from different hosts and different types of ECM fungi are interconnected, and an intricate common hyphal network (common mycorrhizal networks: CMNs) is constructed underground [155, 189, 190]. Once the roots of trees form mycorrhizas with ectomycorrhizal fungi, the whole plant is connected to these CMNs. The absorption area and absorption length of the huge CMNs are much larger than the roots, which can not only promote the absorption and utilization of water and nutrients by host plants but also exchange water and nutrients between different host individuals and even different hosts through CMNs [191–194].

3.2.6 Improving the stress resistance of host plants

3.2.6.1 Enhance the drought resistance of the host

Compared with host trees, ECM fungi have stronger plasticity and ecological adaptability, and most of them have the characteristics of drought and high-temperature resistance. Therefore, under drought conditions, host plants can obtain water from the soil through the mycelium network of ECM fungi to maintain normal growth. At the same time, under drought conditions, ECM can increase the leaf water potential and water retention capacity of host trees and reduce the leaf water saturation deficit value [195, 196]. ECM can also enhance drought resistance by regulating some physiological functions of host plants. For example, ECM fungi enhance host drought resistance by affecting the content or activity of host plant enzymes, including superoxide dismutase (SOD), catalase, peroxidase, and polyphenol oxidase [197–199].

3.2.6.2 Enhance the host's resistance to saline-alkali

Ishida et al. investigated the ectomycorrhizal resources of S.mongolica in saline-alkali land of Northeast China and found that 11 ECM fungi could form ectomycorrhizal with *S. mongolica*, and the dominant species (Geopora sp.) was obvious [200]. A large number of studies have shown that under salt stress conditions, ECM can promote the absorption of water and various nutrients in the soil by tree roots on the one hand, and on the other hand, it can regulate the ion balance in plants and enhance the activity of the host root system, thereby alleviating the damage of saline-alkali stress to the host [201, 202]. In addition, it has been reported that ECM can also improve the respiration and water use efficiency of its host plants to enhance the host's saline-alkali resistance [199].

3.2.6.3 Enhanced host resistance to heavy metals

In recent years, with the excessive development and utilization of mineral resources, heavy metal pollution is becoming more and more serious. These pollutions will cause great toxicity to plants and even cause plant death. Studies have found that mycorrhizal fungi can effectively reduce the toxicity of heavy metals. Huang et al. [203, 204] found that *Pinus massoniana* grown in manganese mining areas can form mycorrhizal fungi with a variety of ECM fungi. Using inoculation experiments, it was found that compared with aseptic seedlings, mycorrhizal seedlings can significantly improve the adaptability of the host, resist heavy metal toxicity, and ensure the biomass growth of the host [203, 204]. Huang et al. [203, 204] studied the effects of growth and heavy metal accumulation and distribution of Pinus tabulaeformis seedlings by planting mycorrhizal seedlings in copper and cadmium-contaminated soil. The results showed that inoculation of mycorrhizal fungi could promote the growth and biomass accumulation of host trees on the one hand, and significantly reduce the accumulation of heavy metals in Pinus tabulaeformis on the other hand, and effectively block the transport of heavy metals from roots to stems and leaves of plants [205, 206]. The current research results show that mycorrhizal fungi can secrete a series of organic acids (such as oxalic acid) or secrete mucus that can chelate heavy metals. These secretions play an important role in the process of plant resistance to heavy metal stress. For example, Ahonen-Jonnarth et al. found that under aluminum stress, ectomycorrhizal seedlings can secrete more oxalic acid than uninoculated seedlings, and oxalic acid can chelate aluminum well [207].

3.2.6.4 Enhance the host's disease resistance

Liu Runjin and Chen Yinglong found that; the sheath and Hatnet in the ECM structure can effectively prevent the mechanical damage of soil pathogens to plant roots [5]. Pine damping-off disease is a common disease on Pinus tabulaeformis seedlings, which causes serious damage to P. tabulaeformis seedlings. Studies have found that inoculation of ectomycorrhizal fungi can significantly prevent P. tabulaeformis damping-off disease [208, 209]. In addition, ectomycorrhizal fungi also secrete some secondary metabolites, such as antibiotics, which can effectively inhibit the invasion of harmful bacteria and fungi in soil to host plants [210].

3.2.6.5 Promote seedling growth and improve the survival rate of afforestation

Afforestation is generally carried out in habitat destruction or difficult site areas. The main reason for the failure of afforestation is the low survival rate of seedlings. A large number of studies have shown that mycorrhizal seedlings can increase the chlorophyll content of the host, increase the photosynthetic rate of the seedlings, and reduce transpiration water loss [211–216]. On the other hand, many ECM fungi can produce plant endogenous hormones. For example, Amanita, Boletus and Rhizopogon can produce indole acetic acid (IAA) and cytokinin, and *R. luteolus* can produce zeatin [217–219]. These hormones can regulate the balance of hormones in plants and promote the growth of host plants, thereby improving the survival rate and preservation rate of afforestation [220].

Author details

YingWu Shi[1,2,3]*, XinXiang Niu[3,4], Hongmei Yang[1,2,3], Ming Chu[1,2,3], Huifang Bao[1,2], Ning Wang[1,2,3], Faqiang Zhan[1,2], Xuanqi Long[1,2], Rong Yang[1,2], Qing Lin[1,2] and Kai Lou[1,2]

1 Institute of Microbiology, Xinjiang Academy of Agricultural Sciences, Urumqi, Xinjiang, China

2 Xinjiang Laboratory of Special Environmental Microbiology, Urumqi, Xinjiang, China

3 Key Laboratory of Agriculture Environment in Northwest Oasis of Agriculture, Urumqi, Xinjiang, China

4 Institute of Soil and Fertilizer and Agricultural Water Conservation, Xinjiang Academy of Agricultural Sciences, Urumqi, Xinjiang, China

*Address all correspondence to: syw1973@126.com

References

[1] Wei SP, Song YJ, Jia LM. Influence of the slope aspect on the ectomycorrhizal fungal community of Quercus variabilis Blume in the middle part of the Taihang Mountains, North China. Journal of Forestry Research. 2021;**32**(01):385-400

[2] Shen HY, Yang BS, Wang H, et al. Changes in soil ectomycorrhizal fungi Community in oak Forests along the Llrban-rural gradient. Forests. 2022;**13**(5):675

[3] Wang JN, Han SJ, Wang CG, et al. Long-tem nitrogen-addition-induced shifts in the ectomycorrhizal fungal community are associated with changes in fine root traits and soil properties in a mixed Pinuskoraiensis forest. European Journal of Soil Biology. 2022;**112**(10):103431

[4] Frank B. On the nutritional dependence of certain trees on root symbiosis with belowground fungi (an English translation of a.B. Frank's classic paper of 1885). Mycorrhiza. 2005;**15**:267-275

[5] Liu RJ, Chen YL. Mycorrhizology. Beijing: Science Press; 2007

[6] Smith SE, Read DJ. Mycorrhizalsymbiosis. London, UK: Academic Press; 2008

[7] Rillig MC. Arbuscular mycorrhizae, glomalin, and soil aggregation. Canadian Journal of Soil Science. 2004;**84**:355-363

[8] Cheng L, Booker FL, Tu C, Burkey KO, Zhou LS, Shew HD, et al. Arbuscular Mycorrhizal fungi increase organic carbon decomposition under elevated C02. Science. 2012;**337**:1084-1087

[9] Clemmensen KE, Bahr A, Ovaskainen O, Dahlberg A, Ekblad A, Wallander H, et al. Roots and associated fungi drive Long-term carbon Sequestrationin boreal Forest. Science. 2013;**339**:1615-1618

[10] Treseder KK, Holden SR. Fungal Carbon Sequestration. Science. 2013;**339**:1528-1529

[11] Averill C, Turner BL, Finzi AC. Mycorrhiza-mediated competition between plants and decomposers drives soil carbon storage. Nature. 2014;**505**:543-545

[12] Wang B, Qiu YL. Phylogenetic distribution and evolution of mycorrhizas in land plants. Mycorrhiza. 2006;**16**:299-363

[13] Parniske M. Arbuscular mycorrhiza: The mother of plant root endosymbioses. Nature Reviews Microbiology. 2008;**6**:763-775

[14] Brundrett M, Bougher N, Dell B, et al. Working with mycorrhizas in forestry and agriculture. [M]: AClAR Monograph. 1996;**32**:30-41

[15] Liu RJ, JiaoH LY, Li M, Zhu XC. Research advances in species diversity of arbuscular mycorrhizal fungi. Chinese Journal of Applied Ecology. 2009;**20**(9):2301-2307

[16] Lin K, Limpens E, Zhang Z, Ivanov S, Saunders DG, et al. Single nucleus genome sequencing reveals high similarity among nuclei of an endomycorrhizal fungus. PLoS Genetics. 2014;**10**:e1004078

[17] Berkeley MJ, Broome CE. Enumeration of the fungi of Ceylon. Journal of the Linnean Society: Botany. 1873;**14**:29-140

[18] Bucholtz F. Beitrage zur kenntnis der Gattung Endogone Link. Beih Zum Botan Centr Abt. 1922;**2**:147-225

[19] Gerdemann JW, Trappe JM. Endogonaceae in the Pacific northwest. Mycologia Memoir. 1974;**5**:1-76

[20] Benjimin RK. Zygomycete and their spores. In: Kendrick B, editor. The Whole Fungus. Vol. I. Ottawa, Canada: National Museums of Canada; 1979. pp. 573-622

[21] Morton JB, Benny GL. Revised classification of arbus-cular mycorrhizal fungi (Zygomycetes): A new order, Glomales, two new suborders, Glominae and Gigaspori-nae, and two families, Acaulosporaceae and Gigaspo-raceae, with an emendation of Glomaceae. Mycotaxon. 1990;**37**:471-491

[22] Almeida RT, Schenck NC. A revision of the genus *Sclerocystis* (Glomaceae, Glomales). Mycologia. 1990;**82**:703-714

[23] Redecker D, Morton JB, Bruns TD. Molecular phylogeny of the arbuscular mycorrhizal fungi *Glomus sinuosum* and *Sclerocystis coremioides*. Mycologia. 2000;**92**:282-285

[24] Schüβler A. Molecular phylogeny, taxonomy, and evolution of Geosiphon pyriformis and arbuscular mycorrhizal fungi. Plant and Soil. 2002;**244**(1):75-83

[25] Oehl F, Silva GAD, Goto BT, Sieverding E. Glomeromycota: Three new genera and glomoid species reorganized. Mycotaxon. 2011;**116**(1):75-120

[26] Opik M, Zobel M, Cantero JJ, Davison J, Facelli JM, Hiiesalu I, et al. Global sampling of plant roots expands the described molecular diversity of arbuscular mycorrhizal fungi. Mycorrhiza. 2013;**23**:411-430

[27] Wang YS, Liu RJ. A checklist of arbuscular mycorrhizal fungi in the recent taxonomic system of Glomeromycota. Mycosystema. 2017;**36**(07):820-850

[28] Liu YP, Sohn B, Wang SZ, Jiang GY, Liu RJ. Advances in the study of genetic diversity of arbuscular mycorrhizal fungi. Biodiversity Science. 2008;**16**(3):225-228

[29] Hijri M, Sanders IR. Low gene copy number shows that arbuscular mycorrhizal fungi inherit genetically different nuclei. Nature. 2005;**433**:160-163

[30] Corradi N, Croll D, Colard A, Kuhn q Ehinger M, Sanders IR. Gene copy number polymorphisms in an arbuscular mycorrhizal fungal population. Applied and Environmental Microbiology. 2007;**73**:366-369

[31] Sanders IR, Alt M, Groppe K, et al. Identification of ribosomal DNA polymorphism among and within spores of the Glomales: Application to studies on the genetic diversity of arbuscular mycorrhizal fungal communities. The New Phytologist. 1995;**130**:419-427

[32] Lloyd-Macgilp SA, Chambers SM, Dodd JC. Diversity of the ribosomal internal transcribed spacers within and among isolates of glomus mosseae and related mycorrhizal fungi. The New Phytologist. 1996;**133**:103-111

[33] Antoniolli ZI, Schachtman DP, Ophel-Keller K, Smith SE. Variation in rDNA ITS sequences in glomus mosseae and Gigaspora margarita spores from a permanent pasture. Mycological Research. 2000;**104**:708-715

[34] Pawlowska TE, Taylor JW. Organization of genetic variation in individuals of arbuscular mycorrhizal fungi. Nature. 2004;**427**:733-737

[35] Kuhn G, Hijri M, Sanders IR. Evidence for the evolution of multiple genomes in arbuscular mycorrhizal fungi. Nature. 2001;**414**:745-748

[36] Bever JD, Morton JB. Heritable variation and mechanisms of inheritance of spore shape within a population of ScutellospoYa pellucida, an arbuscular mycorrhizal fungus. Journal of Ecology. 1999;**86**:1209-1216

[37] Koch AM, Kuhn G, Fontanillas P, Fumagalli L, Goudet G, Sanders IR. High genetic variability and low local diversity in a population of arbuscular mycon-hizal fungi. PNAS. 2004;**101**:2369-2374

[38] Koch AM, Croll D, Sanders IR. Genetic variability in a population of arbuscular mycorrhizal fungi causes variation in plant growth. Ecological Letters. 2006;**9**:103-110

[39] Tawaraya K, Saito M. Effect of vesicular-arbuscular mycorrhizal infection on amino acid composition in roots of onion and white clover. Soil Science and Plant Nutrition. 1994;**40**(2):339-343

[40] Zhang HB, Liang YM, Feng SZ, Zhao ZW, Su YR, He XY. The effects of soil types and plant species on arbuscular mycorrhizal fungi community and colonization in the rhizosphere. Research of Agricultural Modernization. 2016;**37**(01):187-194

[41] Yang XL. Mycorrhizal Diversity in the Larix gmelinii Forest Ecosystems of Da Hinggan Ling Mountains. Hohhot: Inner Mongolia Agricultural University; 2010

[42] Remy W, Taylor TN, Hass H, Kerp H. Four hundred-million-year-old vesicular arbuscular mycorrhizae. Proceedings of the National Academy of Sciences. 1994;**91**:11841-11843

[43] Redecker D, Kodner R, Graham LE. Glomalean fungi from the Ordovician. Science. 2000;**289**:1920-1921

[44] Humphreys CP, Franks PJ, Rees M, Bidartondo MI, Leake JR, Beerling DJ. Mutualistic mycorrhiza-like symbiosis in the most ancient group of land plants. Nature Communications. 2010;**1**:103

[45] Sanders IR. Ecology and evolution of multigenomic arbuscular mycorrhizal fungi. American Naturalist. 2002;**160**:S128-S141

[46] Burggraaf AJP, Beringer JE. Absence of nuclear-DNA synthesis in vesicular Arbuscular Mycorrhizal fungi during Invitro development. New Phytologist. 1989;**111**:25-33

[47] Clapp JP, Rodriguez A, Dodd JC. Inter- and intra-isolate rRNA large subunit variation in Glomus coronatum spores. New Phytologist. 2001;**149**:539-554

[48] Munkvold L, Kjoller R, Vestberg M, Rosendahl S, Jakobsen I. High functional diversity within species of arbuscular mycorrhizal fungi. New Phytologist. 2004;**164**:357-364

[49] van der Heijden MGA, Scheublin TR, Brader A. Taxonomic and functional diversity in arbuscular mycorrhizal fungi is there any relationship? New Phytologist. 2004;**164**:201-204

[50] Mensah JA, Koch AM, Antunes PM, Kiers ET, Hart M, Bucking H. High functional diversity within species of arbuscular mycorrhizal fungi is associated with differences in phosphate and nitrogen uptake and fungal phosphate metabolism. Mycorrhiza. 2015;**25**:533-546

[51] Halary S, Malik SB, Lildhar L, Slamovits CH, Hijri M, Corradi N.

Conserved meiotic machinery in glomus spp., a putatively ancient asexual fungal lineage. Genome Biology and Evolution. 2011;**3**:950-958

[52] Tisserant E, Kohler A, Dozolme-Seddas P, Balestrini R, Benabdellah K, Colard A, et al. The transcriptome of the arbuscular mycorrhizal fungus glomus intraradices (DAOM 197198) reveals functional tradeoffs in an obligate symbiont. New Phytologist. 2012;**193**:755-769

[53] Corradi N, Bonfante P. The arbuscular mycorrhizal symbiosis: Origin and evolution of a beneficial plant infection. PLoS Pathogens. 2012;**8**:e1002600

[54] Tisserant E, Malbreil M, Kuo A, Kohler A, Symeonidi A, Balestrini R, et al. Genome of an arbuscular mycorrhizal fungus provides insight into the oldest plant symbiosis. Proceedings of the National Academy of Sciences of the United States of America. 2013;**110**:20117-20122

[55] Smith SE, Read DJ. Mycorrhizal Symbiosis. Sheffield, UK: Academic Press; 2010

[56] Zhang MQ, Wang YS. The relationship between the distribution of am fungi and environmental factors. Mycosystema. 1999;**18**(1):25-29

[57] Hu JL. The Diversity and Function of Arbuscular Mycorrhizal Fungi Associated with Alfalfa under Different Nutrient Conditions. Beijing: Beijing Forestry University; 2019

[58] Alguacil MM, Tomes MP, Torrecillas E, et al. Plant type differently promotes the arbuscular mycorrhizal fungi biodiversity in their rhizospheres after revegetation of a degraded, semiarid land. In: 19th World Congress of Soil Science, Soil Solutions for a Changing World, Brisbane. 2010

[59] Hiiesalu I, Partel M, Davison J, et al. Species richness of arbuscular mycorrhizal fungi: Associations with grassland plant richness and biomass. New Phytologist. 2014;**203**(1):233-244

[60] Hausmann NT, Hawkes CV. Plant neighborhood control of arbuscular orrhizal community composition. New Phytologist. 2009;**183**(4):1188-1200

[61] Lin X, Feng Y, Zhang H, et al. Long-term balanced fertilization decreases arbuscular mycorrhizal fungal diversity in an arable soil in North China revealed by 454 pyrosequencing. Environmental Science and Technology. 2012;**46**(11):5764-5771

[62] Chen YL, Zhang X, Ye JS, et al. Six-year fertilization modifies the biodiversity of arbuscular mycorrhizal fungi in a temperate steppe in Inner Mongolia. Soil Biology and Biochemistry. 2014;**69**(1):371-381

[63] Liang K, Fan YQ, Kudakwaslie M, et al. The seasonal dynamics of nitrogen and rhizosphere effects in the typical saline-alkali vegetation communities of the Yellow River estuary wetland. Environmental Chemistry. 2019;**38**(10):2327-2335

[64] Camenzind T, Hempel S, Homeier J, et al. Nitrogen and phosphorus additions impact arbuscular mycorrhizal abundance and molecular diversity in a tropical montane forest. Global Change Biology. 2014;**20**(12):3646-3659

[65] Verbruggen E, Xiang D, Chen B, et al. Mycorrhizal fungi associated with high soil N:P ratios are more likely to be lost upon conversion from grasslands to arable agriculture. Soil Biology and Biochemistry. 2015;**86**:1-4

[66] Christinev H, Lainp H, Phil I, et al. Soil temperature affects carbon allocation within arbuscular mycorrhizal networks and carbon transport from plant to fungus. Global Change Biology. 2008;**14**(5):1181-1190

[67] Cai XB, Peng YL, Gai JP. Ecological distribution of arbuscular mycorrhizal fungi in alpine grasslands of Tibet plateau. Chinese Journal of Applied Ecology. 2010;**21**(10):2635-2644

[68] Yin B. The Fungi Survey of Picea Mongolica and Research on Ectomycorrhiza Form Type. Hohhot: Inner Mongolia Agricultural University; 2012

[69] Huang LL, Wan SP, Wang YL, Yu YM, Shi XF, Yu FQ. Mycorrhiza synthesis of Pinus thunbergii with eight ectomycorrhizal fungi species. Acta Edulis Fungi. 2022;**29**(04):98-106

[70] Futai K, Taniguchi T, Kataoka R. Ectomycorrhizae and their importance in Forest cosystems. In: Mycorrhizae: Sustainable Agriculture and Forestry. Netherlands: Springer; 2008. pp. 241-285

[71] Yuam Q. Community Stricture of Ectomycorrhzal Fungi Associate with Fagaceae in a Tropical Montane Rainforest. Haikou: Hainan University; 2016

[72] Wu XT, Zhang B, Qiao LM. Investigation of ectomycorrhizal fungi associated with pine species in Fushun area. Liaoning Forestry Science and Technology. 2018;**06**:61-63

[73] Li M, Gao XH. Community structure and driving factors for rhizosphere ectomycorrhizal fungi of Betula platyphylla in Daqing Mountain. Chinese Journal of Ecology. 2021;**40**(05):1244-1252

[74] Jiang Y, Mo XY, Deng HY, Liu LT. Composition and diversity of ectomycorrhizal fungal community associated with Eucalyptus grandis plantation. Journal of Northwest Forestry University. 2020;**35**(06):153-159

[75] Zou J. Effects of Different Tuber Infected Carya Illinoensis on Plant Growth and Root Microecology. Yaan: Sichuan Agricultural University; 2019

[76] Ma SR, Shi MY, Li Y, He Q, Cui HJ. Research Progress on ectomycorrhizal fungi of Salicaceae. Modern Agricultural Science and Technology. 2020;**10**:100-102

[77] Martin F, Kohler A, Murat C, Veneault-Fourrey C, Hibbett DS. Unearthing the roots of ectomycorrhizal symbioses. Nature Reviews Microbiology. 2016;**14**(12):760-773

[78] Lu LJ, Bai SL, Wang J, Wang TN. Synthesis conditions of Ectomycorrhiza and its formation mechanism. Journal of Microbiology. 2005;**02**:84-87

[79] Gay G, Normand L, Marmeisse R, Sotta B, Debaud JC. Auxin overproducer mutants of Hebeloma cylindrosporum Romagnesi have increased mycorrhizal activity. New Phytologist. 1994;**128**(4):645-657

[80] Feng B, Yang ZL. Ectomycorrhizal symbioses: Diversity of mycobionts and molecular mechanisms that entail the development of ectomycorrhizae. Scientia Sinica(Vitae). 2019;**49**(04):436-444

[81] Lagrange H, Jay-Allgmand C, Lapeyrie F. Rutin, the phenolglycoside from eucalyptus root exudates, stimulates Pisolithus hyphal growth at picomolar concentrations. New Phytologist. 2001;**149**:349-355

[82] Daguerre Y, Plett JM, Veneault-Fourrey C. Signaling pathways driving the development of ectomycorrhizal

symbiosis. In: Martin F, editor. Molecular Mycorrhizal Symbiosis. Hoboken, NJ: John Wiley & Sons, Inc.; 2017. pp. 141-157

[83] Vayssières A, Pěnčík A, Felten J, Kohler A, Ljung K, Martin F, et al. Development of the poplar-Laccaria bicolor Ectomycorrhiza modifies root auxin metabolism, signaling, and response. Plant Physiology. 2015;**169**(1):890-902

[84] Krause K, Henke C, Asiimwe T, Ulbricht A, Klemmer S, Schachtschabel D, et al. Biosynthesis and secretion of Indole-3-acetic acid and its morphological effects on Tricholoma vaccinum-spruce Ectomycorrhiza. Applied and Environmental Microbiology. 2015;**81**(20):7003-7011

[85] Ditengou FA, Müller A, Rosenkranz M, Felten J, Lasok H, van Doorn MM, et al. Volatile signalling by sesquiterpenes from ectomycorrhizal fungi reprogrammes root architecture. Nature Communications. 2015;**6**:6279

[86] Tedersoo L, Smith ME. Ectomycorrhizal fungal lineages: Detection of four new groups and notes on consistent recognition of ectomycorrhizal taxa in high-throughput sequencing studies. In: Tedersoo L, editor. Biogeography of Mycorrhizal Symbiosis. Ecological Studies. Vol. 230. Springer; 2017. pp. 125-142

[87] Tedersoo L, May TW, Smith ME. Ectomycorrhizal lifestyle in fungi: Global diversity, distribution, and evolution of phylogenetic lineages. Mycorrhiza. 2010;**20**(4):217-263

[88] Tedersoo L, Smith ME. Lineages of ectomycorrhizal fungi revisited: Foraging strategies and novel lineages revealed by sequences from belowground. Fungal Biology Reviews. 2013;**27**(3-4):83-99

[89] Tedersoo L, Brundrett MC. Evolution of ectomycorrhizal symbiosis in plants. Biogeography of Mycorrhizal Symbiosis. 2017;**230**:407-467

[90] Feng B, Yang ZL. Ectomycorrhizal symbioses: Diversity of mycobionts and molecular mechanisms that entail the development of ectomycorrhizae. Scientia Sinica (Vitae). 2019;**49**(04):150-158

[91] Meng FR, Shao JW. The ecological distribution of ecto-mycorrhizal fungi in main coniferous forests in Northeast China. Mycosystema. 2001;**20**(3):413-419

[92] Wang Y, Xie ZX. A preliminary survey of of some forest ectomycorrhiza fungi trees in China. Mycosystema. 1983;**2**(1):59-61

[93] Zhao JM, Liu HY, Ni XZ. Study on the classification of Russulaceae in Changbai Mountain. Journal of Jilin Agricultural University. 1998;**20**(1):232-236

[94] Wang SQ, Xu LH. Investigation on ectomycorrhizal fungi resources of main timber species in Northeast China. Liaoning Forestry Science and Technology. 2002;**3**:17-20

[95] Tu LGE, Chen JZ, Wang Y, Fan YG. Macrofungal diversity in broad-leaved Korean pine forest in the Changbaishan National Nature Reserve. Acta Ecologica Sinica. 2010;**30**(17):4549-4558

[96] Wang Q. Ectomycorrhizal community composition of Quercus mongolica. Liaoning Forestry Science and Technology. 2013;**3**:6-9

[97] Song FQ, Liu YK, Du CM, Yang FS. Diversity of ectomycorrhizal macro-fungi at Xiao Hinggan Mountain and medium optimization for Suillus grevillei using response surface methodology.

Journal of Natural Resources. 2011;**26**(3):440-449

[98] Lang QL, Zhang XQ, Wang FT, Gang SB, Wang H. A preliminary report on the investigation of tussah mycorrhizal fungi resources and the observation of fruiting body occurrence period in Fengcheng area of Liaoning Province. North Sericulture. 2003;**24**(96):31-35

[99] Wang H, Dai LM, Shao GF, Lang QL, Yang BS, Deng HB, et al. Ecological distribution of mycorrihizal fungi associated with oak in Dandong district of Liaoning Province. Chinese Journal of Applied Ecology. 2003;**14**(12):2149-2152

[100] Cui L, Mu LQ. The morphological and anatomical characteristics of ectomycorrhizae colonizing Tilia amurensis Rupr. in Heilongjiang Province. Chinese Journal of Ecology. 2014;**33**(9):2490-2500

[101] Tang M, Chen H, Guo JL, Hou DW. A study of the ectomycorrhiza of poplar. Scientia Silvae Sinicae. 1994;**30**(5):437-441

[102] Tian ML. Investigation and distribution of Russulaceae resources in Gansu Province. Journal of Jilin Agricultural University. 1998;**20**(S1):219-223

[103] Wu JK, He LL, Li JL. A primary study on macrofungi Resouces from Shaanxi. Journal of Northwest University(Natural Science Edition). 1996;**26**(5):435-438

[104] Wu CH, Wang JR, Yang JX, Wu GH. A study on ectomycorrhiza and mycorrhiza fungi in Taibai Mountain nature reserve. Journal of Northwest Sci-Tech University of Agriculture and Forestry (Natural Science Edition). 2001;**29**(2):56-60

[105] Pu X. The ectomycorrhizal macromycetes resources inLonglan area of Gansu. Journal of Gansu Forestry Science and Technology. 2000;**25**(2):1-5

[106] Xun LH. Studies on Diversity and Nutrition Constituent of Macrofungi in Helan Mountain in Ningxia. Inner Mongolia: Inner Mongolia Agricultural University; 2012

[107] Zhang RQ, Tang M, Zhang HH, Du XG. A preliminary investigation on ectomycorrhizal fungi in Huoditang Forest region of Qinling Mountains. Journal of Qingdao Agricultural University(Natural Science). 2011;**28**(3):168-173

[108] Qi P, Li JZ, Li AL, Dai L, Ding SG. Preliminary survey report of macro-fungi germplasm resources in the middle area of Qinling Mountains. Edible Fungi of China. 2013;**32**(1):8-13

[109] Ding XW, Liu KH, Deng BW, Chen WQ. Preliminary survey of macro fungi resources in Zibai Mountain of Qinling area. Guizhou Agricultural Sciences. 2012;**40**(10):111-113

[110] Guli A, Feng L, Qing XZ, Chen J, Marhaba, Yang XP, et al. Investigation on wild fungi resources in Habahe plain of Altai prefecture (I). Xinjiang Agricultural Sciences. 2015;**52**(9):1707-1714

[111] Chen ZH, Zhang ZG, Zhang P, Zhang TX, Zhang ZH, Xiao QY. The resources, taxonomy and ecological features of amanita in mount. Mang, Hunan. Edible Fungi of China. 1997;**16**(6):27-29

[112] Li JZ. Two new records of am Anita in China. Acta Mycologica Sinica. 1996;**15**(2):154-156

[113] Cui B, Ma J, Li LC. Study on the macrofungus resources of russulaceae

family in Henan (II). Henan Science. 1998;**16**(3):323-329

[114] Zhang LP, Tao AJ, Zhang Y, Luang FG, Wang S, Ling CY. The ecological distribution and resource assessment of macrofungi in Tongboshan natural Reserve of Jiangxi Province. Acta Agriculturae Universitatis Jiangxiensis. 2013;**35**(6):1229-1235

[115] Li JZ, Chen SM, Dan XQ. Macrofungus resources of the Shunhuang Mountain natural protection region. Life Science Research. 2006;**10**(3):276-281

[116] Li JZ, Yu XL, Zhou JL, Tang M, Wang X, Chen Q, et al. The macrofungus resources investigation and the ecologic distribution and economy meaning of the Nan-Yue Mountain natural protection region. Life Science Research. 2007;**03**:218-226

[117] Huang J, Nara K, Lian CL, Zong K, Peng KJ, Xue SG, et al. Ectomycorrhizal fungal communities associated with Masson pine (Pinus massoniana lamb.) in Pb-Zn mine sites of central South China. Mycorrhiza. 2012;**22**(8):589-602

[118] Qin L, Xv J, Ma X. Research on Symbiotical fungi species and Ectomycorrhizae occurrence of chestnut (Castanea mollissima BL.). Journal of Beijing University of Agriculture. 1995;**10**(1):71-76

[119] Cai HS, Liu HX, Guo YM, Cui GF. Brief report on investigation of macro-fungi in Beijing Song mountain nature reserve. Ecological Science. 2003;**22**(3):250-251

[120] Tian HM, Liu TZ, Duan YP. Resources and economic value of the fungal family Russulaceae in Chifeng, Inner Mongolia, China. Journal of Fungal Research. 2014;**12**(2):83-87

[121] Fan YJ, Yan W. Morphological type and molecular identification of Ectomycorrhizae on Betula platyphyllain Inner Mongolia area. Acta Botanica Boreali-Occidentalia Sinica. 2013;**33**(11):2209-2215

[122] Bai SL, Yan W, Ma RH, Wang TN. Investigation of ECM fungi resources in Mt. Daqing and Mt. Manhan. Mountain Research. 2001;**19**(1):44-47

[123] Fan YJ, Yan W, Wang LY. Morphological type and molecular identification of Ectomycorrhizae on Picea crassifolia in Helan Mountain. Scientia Silvae Sinicae. 2011;**47**(6):108-113

[124] Bai SL, Liu Y, Zhou J, Dong Z, Fan R. Resources investigation and ecological study on ectomycorrhizal fungi in Daqingshan Mountains, Inner Mongolia. Acta Ecologica Sinica. 2006;**26**(3):837-841

[125] Gong MQ, Chen Y. A study on Pinus and eucalyptus Ectomycorrhiza in South China. Forest Research. 1991;**4**(3):323-327

[126] Rao BQ. Study on mycorrhizal community of Pinus massoniana lamb. In: Wuyishan Nature Reserve. Xiamen: Xiamen University; 2000

[127] Qian XM, Huang YJ, Zhang YH, Wu JL, Chen RH, Liu GF, et al. Biodiversity of the ectomycorrhiza on the rare and endangered tree species Tuga chinensis tchekiangensis(Flous) Cheng in Wuyishan nature reserve. Journal of Fujian Agriculture and Forestry University: Natural Science Edition. 2007;**36**(2):180-185

[128] Li HB, Wu XQ, Wei HL, Fu LZ, Wu QQ, Chen YL, et al. Brief report on investigation of ectomycorrhizal fungi resources in Lishui area, Zhejiang Province. Edible Fungi of China. 2005;**24**(5):10-14

[129] Ke LX, Liu BR. Resource and ecological distribution of ectomycorrhizal fungi under pine forests of Huangshan Mountain district. Chinese Journal of Applied Ecology. 2005;**16**(3):455-458

[130] Chen TX, Chen SL. Species composition and ecological characteristics of macrofungi in Zijin Mountain, Nanjing. Journal of Ecology and Rural Environment. 2012;**28**(4):373-379

[131] Bi GC, Zang M, Guo XZ. Distribution of ectomycorrhizal fungi under several chief forest types in alpine coniferou region of northwestern Yunnan. Scientia Silvae Sinicae. 1989;**25**(1):33-39

[132] He SC. A preliminary study on the resources of forest ectotrophic fungi and their ecological distribution from Guizhou province, China. Guizhou Science. 1991;**01**:51-58

[133] Tan FH, Wang YZ. Investigation on species of ectomycorrhizal fungi of pine and eucalyptus in Sichuan. Journal of Sichuan Forestry Science and Technology. 2000;**21**(3):65-69

[134] Yu FQ, Xiao YQ, Liu PG. Spatiotemporal distribution of ectomycorrhizal fungi in Pinus yunnanensis forests. Acta Ecologica Sinica. 2007;**27**(6):2325-2333

[135] Zhu TH, Zhang J, Hu TX, Li XW, Liu YG. The study on ectomycorrhizal fungi associated with eucalyptus in Sichuan. Journal of Sichuan Agricultural University. 2001;**19**(2):137-140

[136] Liu XJ, Xv AS, Deng LJ. The resources of Russulaceae in Sejila Mountain of Xizang. Edible Fungi of China. 2010;**29**(4):8-9

[137] Zhou GL, Deng HP, Zhang JH. The macrofungi resources diversity of Jinyunshan natural reserve. Guizhou Agricultural Sciences. 2012;**40**(12):126-130

[138] Luo GT, Wang JX, Yuan W. Investigation of the macrofungi of Ectotrophic mycorrhiza of Pinus massoniana in Kaili City, Guizhou Province. Edible Fungi of China. 2007;**26**(2):12-14

[139] Rinaldi AC, Comandini O, Kuyper TW. Ectomycorrhizal fungal diversity: Seperating the wheat from the chaff. Fungal Diversity. 2008;**33**:1-45

[140] Tedersoo L, Sadam A, Zambrano M, et al. Low diversity and high host preference of ectomycorrhizal fungi in Western Amazonia, a neotropical biodiversity hotspot. The ISME Journal. 2010;**4**(4):465-471

[141] Wurzburger N, Hartshorn AS, Hendrick RL. Ectomycorrhizal fungal community structure across a bog-forest ecotone in southeastern Alaska. Mycorrhiza. 2004;**14**(6):383-389

[142] Smith ME, Henkel TW, Aime MC, et al. Ectomycorrhizal fungal diversity and community structure on three co-occurring leguminous canopy tree species in a Neotropical rainforest. New Phytologist. 2011;**192**(3):699-712

[143] Lamit LJ, Holeski LM, Flores-Rentería L, et al. Tree genotype influences ectomycorrhizal fungal community structure: Ecological and evolutionary implications. Fungal Ecology. 2016;**24**:124-134

[144] Bonito G, Brenneman T, Vilgalys R. Ectomycorrhizal fungal diversity in orchards of cultivated pecan (Carya illinoinensis; Juglandaceae). Mycorrhiza. 2011;**21**(7):601-612

[145] Ji RQ, Gao TT, Li GL, Xv Y, Xing PJ, Zhou JJ, et al. Correlation between

ectomycorrhizal fungal community and environmental factors in Pinus koraiensis forest in Northeast China. Mycosystema. 2020;**39**(04):743-754

[146] Zhao M, Hao LF, Zhang M, Teng H, Yan HX, Bai SL. Ectomycorrhizal fungal community structure characteristics of natural Pinus sylvestris var. mongolica with different age classes in Honghuaerji, greater Khingan Mountains. Mycosystema. 2019;**38**(9):1420-1429

[147] Yang Y. Ectomycorrhizal Fungal Community in the Rhizospheric Soil of Eight Tree Species in Inner Mongolia. Hohhot: Inner Mongolia Agricultural University; 2018

[148] Long D. Diversity Maintenance Mechanism of Ectomycorrhizal Fungal Community Associated with *Pinus tabuliformis* in North China. Yangling: Northwest A&F University; 2017

[149] Wang XB. Research on the Ectomycorrhizal Fungal Community Associated with Major Oaks at Qinling Mountains and Loess Plateau. Yangling: Northwest A&F University; 2017

[150] Zhang TT, Geng ZC, Xv CY, Zhang XP, Du C, Wang ZK, et al. Diversity of ectomycorrhizal fungi associated with Larixg melinii in Xinjiashan forest region of Qinling Mountains. Acta Microbiologica Sinica. 2018;**58**(3):443-454

[151] Noack F. Ueber mykorhizenbildende Pilze. Bot Zeit. 1889;**24**:391-404

[152] Melin E. Über die mykorrhizenpilze von Pinus silvestris L. und Picea abies (L.) karst. Svensk Botanisk Tidskrift. 1921;**15**:192-203

[153] Melin E. Experimentelle Untersuchungen über die Konstitution und Ökologie der Mykorrhizen von Pinus silvestris L. und Picea Abies (L.) Karst. Mykol Untersuch Ber. 1923;**2**:73-331

[154] Molina R, Massicotte H, Trappe JM. Specificity phenomena in mycorrhizal symbioses: Community-ecological consequences and practical implications. In: Allen MF, editor. Mycorrhizal Functioning, an Integrative Plant–Fungal Process. New York, NY: Springer; 1992. pp. 357-423

[155] Taylor AF, Alexander IAN. The ectomycorrhizal symbiosis: Life in the real world. Mycologist. 2005;**19**(3):102-112

[156] Hibbett DS, Binder M, Bischoff JF, Blackwell M, Cannon PF, Eriksson OE, et al. A higher-level phylogenetic classification of the fungi. Mycological Research. 2007;**111**(5):509-547

[157] Hibbett DS, Luz-Beatriz G, Donoghue MJ. Evolutionary instability of ectomycorrhizal symbioses in basidiomycetes. Nature. 2000;**407**(6803):506

[158] Chen LQ. Studies on symbiotic mycorrhiza fungi with masson pine. Forest Research. 1989;**04**:357-362

[159] Zhao Z, Ma KX, Duan AA. Research on types and ecological characteristics of ectomycorrhiza of *Populus tomentosa*. Scientia Silvae Sinicae. 1993;**01**:12-18

[160] Bai SL. Study on Distribution and Selection of Ectomycorrhizal Fungi in Daqingshan Mountain, Inner Mongolia. Beijing: Beijing Forestry University; 2006

[161] Fan YJ. Morphologic Analysis and Molecular Identification of Ectomycorrhizae of Four Dominant Tree Species in Inner Mongolia Area. Hohhot: Inner Mongolia Agricultural University; 2009

[162] Zhang WQ. The Study on the Diversity of Ectomycorrhizal Fungi of Pinus sylvestris Var. Mongolica Litv and Mycorrhizal Biotechnology. Hohhot: Inner Mongolia Agricultural University; 2013

[163] Wang Q, Gao C, Guo LD. Ectomycorrhizae associated with Castanopsis fargesii (Fagaceae) in a subtropical forest, China. Mycological Progress. 2011;**10**:323-332

[164] Wang Q, He XH, Guo LD. Ectomycorrhizal fungus communities of Quercus liaotungensis Koidz of different ages in a northern China temperate forest. Mycorrhiza. 2012;**22**(6):461-470

[165] Ding Q, Liang Y, Legendre P, He XH, Pei KQ, Du XJ, et al. Diversity and composition of ectomycorrhizal community on seedling roots: The role of host preference and soil origin. Mycorrhiza. 2011;**21**(8):669-680

[166] Yin RH, Yan W, Wei J. Ectomycorrhizal diversity of Picea Crassifolia in Helan Mountain of Inner Mongolia. Inner Mongolia Forestry Investigation and Design. 2017;**40**(2): 71-72, 75

[167] Geng R, Geng ZC, Huang J, He WX, Hou L, She D, et al. Diversity of ectomycorrhizal fungi associated with Quercus aliena in Xinjiashan forest region of Qinling Mountains. Mycosystema. 2016;**07**:833-847

[168] Geng R, Geng ZC, Huang J, He WX, Hou L, She D, et al. Diversity of ectomycorrhizal fungi associated with Picea asperata in Xin Jiashan Forest of Qinling Mountains. Acta Microbiologica Sinica. 2015;**07**:905-915

[169] Gao C, Zhang Y, Shi NN, Zheng Y, Chen L, Wubet T, et al. Community assembly of ectomycorrhizal fungi along a subtropical secondary forest succession. New Phytologist. 2015;**205**(2):771-785

[170] Wen ZG. Ectomycorrhizal Fungal Communities on the Endangered Chinese Douglas-Fir (Pseudotsuga Sinensis) and Identification of Rhizopogo n-Pseudots Ectomycorrhizal Symbiosis. Nanjing: Nanjing Agricultural University; 2014

[171] Han Q, Huang J, Long D, Wang X, Liu J. Diversity and community structure of ectomycorrhizal fungi associated with Larix chinensis across the alpine treeline ecotone of Taibai Mountain. Mycorrhiza. 2017;**2**:1-11

[172] Wang X, Liu J, Long D, Han Q, Huang J. The ectomycorrhizal fungal communities associated with quercus liaotungensis, in different habitats across northern China. Mycorrhiza. 2017;**27**(5):441-449

[173] van der Heijden MGA, Klironomos JN, Ursic M, Moutoglis P, Streitwolf-Engel R, Boller T, et al. Mycorrhizal fungal diversity determines plant biodiversity, ecosystem variability and productivity. Nature. 1998;**396**(6706):69-72

[174] Wang D, Zhou H, Zuo J, Chen P, She Y, Yao B, et al. Responses of soil microbial metabolic activity and community structure to different degraded and restored grassland gradients of the Tibetan plateau. Frontiers in Plant Science. 2022;**13**:770315

[175] Luo QY, Wang XJ, Lin SS, Li YY, Sun L, Jin L. Mechanism and application of bioremediation to heavy metal polluted soil using arbuscular mycorrhizal fungi. Acta Ecologica Sinica. 2013;**33**(13):3898-3906

[176] Zu YQ, Lu X, Zhan FD, Hu WY, Li Y. A review on roles and mechanisms

of Arbuscular Mycorrhizal fungi in phytoremediation of heavy metals–polluted soils. Plant Physiology Journal. 2015;**51**(10):1538-1548

[177] Wang HJ, Bao YY, Niu TX, Li XF. Effect of combined application of microbial inoculum and fertilizer on fertility of open-pit mine dump reclaimed soil. Acta Agriculturae Boreali-Sinica. 2014;**29**(04):186-191

[178] Zhang ZF, Zhang JC, Huang YQ, Xv GP, Zhang DN, Yu YC. Effects of mycorrhizal fungi on the drought tolerance of *Cyclobalanopsis glauca* seedlings. Acta Ecologica Sinica. 2016;**36**(11):3402-3410

[179] Hayman DS. Influence of soils and fertility on activity and survival of vesiculararbuscular mycorrhizal fungi. Phytopathology. 1982;**72**:1119-1125

[180] Ames RN, Reid CP, Potter LK, et al. Hyphal uptake and transport of nitrogen from two ^{15}N-labelled sources by glomus mosseae, a vesicular-arbuscular mycorrhizal fungus. New Phytologist. 1993;**95**:381-396

[181] Harrison MJ, Van Buuren ML. A phosphate transporter from the mycorrhizal fungus glomus versif Orme. Nature. 1995;**378**:626-629

[182] Dimond JM. Colonization of a volcano inside a volcano. Nature. 1977;**270**:13-14

[183] Trappe JM. Phylogenetic and ecological aspects of mycotrophy in the angiosperms from an evolutionary standpoint. In: Safer GR, editor. Ecophysiology of VA Mycorrhiz al Plants. Florida: CRC Press; 1987. pp. 5-26

[184] Reeves FB, Wagner D, Moorman T, Kiel J. The role of endomycorrhizae in revegetation practices in semiarid west. I. A comparison of incidence of mycorrhizae in severely disturbed v/s natural environments. American Journal of Botany. 1979;**66**:6-13

[185] Read D. The ties that bind. Nature. 1997;**388**:517-518

[186] Mosse B. Advances in the study of vesicular-arbuscular mycorrhiza. Annual Review of Phytopathology. 1973;**11**:171-196

[187] Grime JP, Mackey JML, Hillier SH, Read DJ. Floristic diversity in a model system using experimental microcosms. Nature. 1987;**328**:420-422

[188] Bergelson JM, Crawley MJ. Mycorrhizal infection and plant species diversity. Nature. 1988;**334**:202

[189] Booth MG. Mycorrhizal networks mediate overstorey-understorey competition in a temperate forest. Ecology Letters. 2004;**7**(7):538-546

[190] Nara K. Ectomycorrhizal networks and seedling establishment during early primary succession. New Phytologist. 2006;**169**(1):169-178

[191] He X, Bledsoe CS, Zasoski RJ, Southworth D, Horwath WR. Rapid nitrogen transfer from ectomycorrhizal pines to adjacent ectomycorrhizal and arbuscular mycorrhizal plants in a California oak woodland. New Phytologist. 2006;**170**(1):143-151

[192] He X, Xu M, Qiu GY, Zhou J. Use of 15N stable isotope to quantify nitrogen transfer between mycorrhizal plants. Journal of Plant Ecology. 2009;**2**(3):107-118

[193] Bai SL, Li GL, Liu Y, Dumroese RK, Lv RH. Ostryopsis davidiana seedlings inoculated with ectomycorrhizal fungi facilitate

formation of mycorrhizae on *Pinus tabulaeformis* seedlings. Mycorrhiza. 2009;**19**(6):425-434

[194] Walder F, Boller T, Wiemken A, Courty PE. Regulation of plants' phosphate uptake in common mycorrhizal networks: Role of intraradical fungal phosphate transporters. Plant Signaling & Behavior. 2016;**11**(2):e1131372

[195] Wu BY, Liang NK. Effect of Ectomycorrhizae on the drought resistance of Pinus tabulae formis seedlings. Journal of Beijing Forestry University. 1991;**S3**:281-290

[196] Lv Q, Lei ZP. A study on effect of Ectomycorrhizae on promoting Castanea mollissima resistance to drought and its mechanism. Forest Research. 2000;**03**:249-256

[197] Chen H, Tang M, Liu XD, Li ZF. Effects of ectomycorrhizal fungi on poplar resistance to canker disease. Acta Phytopathologica Sinica. 1996;**04**:83

[198] Osonubi O, Okon IE, Bamiduro TA. Effect of different fungal inoculation periods on performance of Gmelina seedlings under dry soil conditions. Forest Ecology and Management. 1990;**37**(4):223-232

[199] Ruiz-Lozano JM, Azcón R, Palma JM. Superoxide dismutase activity in arbuscular mycorrhizal Lactuca sativa plants subjected to drought stress. New Phytologist. 1996;**134**(2):327-333

[200] Ishida TA, Nara K, Ma S, Takano T, Liu S. Ectomycorrhizal fungal community in alkaline-saline soil in northeastern China. Mycorrhiza. 2009;**19**(5):329-335

[201] Huang Y, Jiang XY, Liang ZC, Li T. Effect of ectomycorrhizal fungi on growth and physiology of Pinus tabulaeformis seedlings under saline stress. Journal of Agro-Environment Science. 2006;**25**(6):1475-1480

[202] Kernaghan G, Hambling B, Fung M, Khasa D. In vitro selection of boreal ectomycorrhizal fungi for use in reclamation of saline-alkaline habitats. Restoration Ecology. 2002;**10**(1):43-51

[203] Huang J, Nara K, Zong K, Lian C. Soil propagule banks of ectomycorrhizal fungi along forest development stages after mining. Microbial Ecology. 2015;**69**(4):768-777

[204] Huang J, Nara K, Zong K, Wang J, Xue S, Peng K, et al. Ectomycorrhizal fungal communities associated with Masson pine (Pinus massoniana) and white oak (Quercus fabri) in a manganese mining region in Hunan Province, China. Fungal Ecology. 2014;**9**(1):1-10

[205] Huang Y, Huang ZJ. Ectomycorrhizae and heavy metals resistance of higher plants. Chinese Journal of Ecology. 2005;**04**:422-427

[206] Huang Y, Peng B, Li T, Liang ZC. Growth and element accumulation of pinus tabulaeformis seedlings influenced by inoculation of ectomycorrhizal fungi in cu and cd contaminated soil. Chinese Journal of Plant Ecology. 2007;**05**:923-929

[207] Ahonen-Jonnarth ULLA, Van Hees PAW, Lundström US, Finlay RD. Production of organic acids by mycorrhizal and non-mycorrhizal Pinus sylvestris L. seedlings exposed to elevated concentrations of aluminium and heavy metals. New Phytologist. 2000;**146**:557-567

[208] Guo XZ, Tan F, He SL. Ectomycorrhizal fungi control damping-off disease of *Pinus tabulaeformis* seedlings. Forestry Science & Technology. 1981;**02**:23-25

[209] Zhang RQ, Tang M, Zhang HH. Effects of four ectomycorrhizal species on controlling damping off and promoting the growth of *Pinus tabulaeformis* seedlings. Mycosystema. 2011;**05**:812-816

[210] Lei ZP, Jing JR, Wang CW. Antagonism of ectomycorrhizal fungi to pathogens of plant roots. Scientia Silvae Sinicae. 1989;**06**:502-508

[211] Yan XF, Wang Q. Effects of co-inoculation with two ectomycorrhizal fungi on quercus liaotungensis seedlings. Chinese Journal of Plant Ecology. 2004;**01**:17-23

[212] Yan XF, Wang Q. Effects of ectomycorrhizal inoculation on the seedling growth of *Quercus liaotungensis*. Chinese Journal of Plant Ecology. 2002;**06**:701-707

[213] Wang Y, Ding GJ. Effects of ectomycorrhizal on growth, physiological characteristics and nutrition in *Pinus massoniana* seedlings. Journal of Nanjing Forestry University: Natural Sciences Edition. 2013;**02**:97-102

[214] Wang Y, Ding GJ. Effects of exogenous mycorrhiza on growth of *Pinus massoniana* seedlings. Journal of Central South University of Forestry & Technology. 2011;**04**:74-78

[215] Liu XG, Guo SJ, Zhao H, Ma LY. Effect of different ectomycorrhizal fungi on photosynthesis and transpiration characteristics of *Quercus variabilis* seedlings. Journal of Zhejiang Forestry Science and Technology. 2010;**06**:1-5

[216] Gao Y, Wu XQ. Effects of several ectomycorrhizal fungi on the chlorophyll content and chlorophyll *fluorescence parameters* in different pine seedlings. Journal of Nanjing Forestry University: Natural Sciences Edition. 2010;**06**:9-12

[217] Durand DJC, Casselton LA, Gay G. Isolation and preliminary characterization of 5-fluoroindole-resistant and IAA-overproducer mutants of the ectomycorrhizal fungus Hebeloma cylindrosporum Romagnesi. New Phytologist. 1992;**121**(4):545-553

[218] Kraigher H, Grayling A, Wang TL, Hanke DE. Cytokinin production by two ectomycorrhizal fungi in liquid culture. Phytochemistry. 1991;**30**(7):2249-2254

[219] Ulrich JM. Auxin production by mycorrhizal fungi. Physiologia Plantarum. 1960;**13**(3):429-443

[220] Yu F, Liu P. Reviews and prospects of the ectomycorrhizal research and application. Acta Ecologica Sinica. 2001;**22**(12):2217-2226

Section 3

Ectomycorrhizal Fungi

Chapter 5

Ectomycorrhizal Inoculum: A Key Tool for Rehabilitation of Natural Forests

Sana Jabeen

Abstract

Deforestation is among the greatest challenges the Earth is facing. The annual deforestation rate is more than 3%. To uplift the economic growth of any country, the forest cover should be at least 25%. To overcome this problem, rapid afforestation and reforestation strategies are required. Inoculation of ectomycorrhizal fungi growing efficiently in the biodiversity-rich regions could be a leading approach in this regard. Several ectomycorrhizal fungi have been reported in association with many coniferous and deciduous tree species growing in these regions. The success of this association is mainly based on the mutual exchange of nutrients between the symbionts. These ectomycorrhizal fungi can mitigate the stress conditions and enhance the seedling survival. Inoculation of these fungi with indigenous tree species of the region can greatly improve plant growth and survival. This symbiosis may play a major role in the function, maintenance, and evolution of biodiversity and ecosystem stability and productivity.

Keywords: *Ascomycota*, *Basidiomycota*, *Mucoromycota*, mutualistic symbiotic, temperate, tropical

1. Introduction

Deforestation is indeed one of the greatest challenges that the Earth is facing, and has several detrimental impacts. The removal of forests results in the destruction of diverse habitats that support a wide range of plant and animal species. Many species, including those unique to specific ecosystems, may face extinction as their habitats disappear. Even if some areas remain untouched, the fragmentation of forests into smaller patches can isolate populations, limit genetic diversity, and make it difficult for species to survive and reproduce.

Deforestation releases stored carbon into the atmosphere, contributing to global warming. It disrupts climate patterns, impacting agriculture, water resources, and settlements. Forests play a crucial role in water cycle regulation, and deforestation leads to altered river flow and increased flooding. The removal of trees weakens soil structure, making areas prone to erosion and landslides, especially in hilly regions. Preserving forests is vital to mitigate these environmental challenges. Many indigenous communities rely on forests for their livelihoods, obtaining food, medicine, and materials for shelter and cultural practices. Deforestation can displace these communities, disrupt their traditional way of life, and lead to the loss of cultural and ecological knowledge. While deforestation may provide short-term economic benefits through activities like logging and conversion of land for agriculture, the long-term consequences can be severe.

Loss of ecosystem services, including clean water, pollination, and climate regulation, can have significant economic impacts. Planting new trees to replace those that have been cut down helps restore ecosystems and mitigate the impacts of deforestation. Reforestation projects are essential for combating climate change and preserving biodiversity. Implementing practices that ensure the responsible use of forest resources, such as selective logging and regeneration plans, can help maintain the ecological integrity of forests while meeting human needs for wood and other products. Supporting and promoting alternative livelihoods for communities dependent on forest resources can reduce pressure on forests. This may involve the development of sustainable agroforestry, ecotourism, or non-extractive economic activities.

An annual deforestation rate exceeds 3%. To uplift the economic growth of any country, the forest cover should be at least 25%. Addressing deforestation requires a holistic approach that considers environmental, social, and economic factors. International collaboration and policies that promote sustainable land use practices are crucial in mitigating the negative impacts of deforestation and promoting a more harmonious relationship between humans and the environment. Rapid afforestation and reforestation strategies are essential and effective remedies to address deforestation and increase forest cover.

Ectomycorrhiza is a type of mutualistic symbiotic association between certain groups of fungi and the roots of higher plants. In this association, the fungi do not penetrate the root cells; instead, they form a mycelium between the cortical cells. This mycelium is known as Hartig's net. This relationship is beneficial for both partners. The fungus enhances the plant's nutrient absorption capabilities, especially for minerals like phosphorus and nitrogen, which can be less accessible to the plant in the soil. In return, the plant provides the fungus with carbohydrates produced through photosynthesis. The fungal hyphae form a sheath around the outer surface of the plant root tips, creating a structure known as a "mantle". The mantle layer offers protection against certain soil-borne pathogens, helping to enhance the plant's resistance to diseases and helps plants in coping with stress. The hyphae which originate from the mantle layer extend into the soil, forming a dense network around the plant roots to play a role in improving water uptake by the host plant, especially in conditions of water stress [1–3].

2. Evolution of ectomycorrhiza

Numerous studies propose that diverse ectomycorrhizal associations evolved independently, occurring at least once in the *Pinaceae* and in various lineages of angiosperms [4, 5]. However, the geographic origins and subsequent spread of ectomycorrhizal associations are still unclear. Since ectomycorrhizas are predominantly found in boreal and temperate forests today [6], examined whether these associations originated in these environments and subsequently spread to tropical regions, or if ectomycorrhizas emerged independently in tropical regions. Until recently, the sole fossil evidence for ectomycorrhizas came from the roots of Eocene *Pinaceae* on Vancouver Island [7], potentially indicating a northern latitude origin for ectomycorrhizas.

3. Plants forming ectomycorrhizal association

Approximately 8000 species of seed plants, which represent about 3% of all seed plant species, engage in this type of symbiosis [8, 9]. Even though being a minority among seed plants, these ectomycorrhizal (ECM) species are incredibly significant ecologically and economically. This is because they play a dominant role in forest and woodland ecosystems across much of the Earth's surface. Their presence shapes these ecosystems and influences the diversity and abundance of other organisms living within them [4]. Despite the prevalence of woody perennials, some non-woody plants also form ectomycorrhizal associations. For example, certain sedges (plants resembling grasses with triangular stems) e.g., *Kobresia* spp., and herbaceous species in the *Polygonum* genus (which includes various types of flowering plants), can engage in ectomycorrhizal symbiosis. This expands the diversity of ECM plant species beyond woody perennials [10].

It is believed that ectomycorrhizal associations are primarily found in temperate and boreal forests, with their presence elsewhere being sporadic and of lesser ecological significance. It is certainly true that dominant tree species in these regions, such as those from the *Fagaceae*, *Betulaceae*, *Salicaceae*, *Myrtacea, Leptospermoideae*, and *Pinaceae* families, typically form ectomycorrhizal associations in their natural habitats. Furthermore, the ectomycorrhizal habit exhibits specific adaptations that enhance nutrient uptake in temperate and boreal forests, as documented by [11].

However, contrary to this belief, extensive portions of land beyond temperate and boreal zones also host vegetation with a significant ectomycorrhizal component. For instance, Arctic and alpine habitats in the northern hemisphere are characterized by dwarf shrub communities dominated by species such as Dryas and Salix, which form ectomycorrhizal associations supporting diverse communities of fungal partners. Similarly, regions with winter-wet ecosystems, such as those found in the Mediterranean region, exhibit a pronounced presence of ectomycorrhizal associations among species like *Pinus*, *Cistus*, *Arbutus*, and *Arctostaphylos*. This demonstrates that ectomycorrhizal associations are not confined to temperate and boreal forests but are prevalent across various ecosystems, highlighting their broader ecological importance and distribution [12].

However, it is in the tropics that the occurrence and importance of ECM host species have been most consistently underestimated. Consider the *Dipterocarpaceae*, of which all members form ectomycorrhizas. This diverse family of over 500 spp. is the source of most tropical hardwood timber and is of immense economic and ecological importance.

Sarcolaenaceae, sharing a common ancestor with *Dipterocarpaceae*, has been found to be ectomycorrhizal by [13], suggesting that the ectomycorrhizal habit in this clade was present more than 88 million years ago. The discovery of Pakaraimea being ECM pushes this origin back to 130 million years ago. Understanding ECM fungi in dipterocarps could offer insights into ECM evolution and tropical forest ecology.

Ectomycorrhizas are prevalent in certain members of *Fabaceae* primarily in African tropical rainforests and savannah woodlands. Genera such as *Tetraberlinia*, *Microberlinia*, and *Julbernardia* dominate rainforests, while *Isoberlinia* and *Brachystegia* cover vast areas in African woodlands. These ecosystems support diverse ectomycorrhizal fungi [14–17].

Two temperate ectomycorrhizal plant taxa; *Fagaceae* and *Pinaceae*, have radiated into natural tropical forests. Tropical oak species like *Lithocarpus*, *Castanopsis*, and *Trigonobalanus* occur in Central America and SE Asia along with ectomycorrhizal dipterocarps.

Endemic ectomycorrhizal tropical pines like *Pinus oocarpa*, *P. patula*, and *P. caribea* are found in Mexico, Central America, the Caribbean, and SE Asia. The ectomycorrhizal fungi of these natural tropical pine forests are not well studied. However, several tropical pines have been extensively used in plantations throughout the tropics outside their natural range, and often in areas that have not previously supported ectomycorrhizal vegetation. Ironically, in these plantations, the mycobionts are either temperate species introduced with inoculum from Europe, or the ubiquitous ectomycorrhizal genus *Pisolithus* [18]. *Eucalyptus* spp. are widespread throughout temperate and subtropical forest ecosystems in Australia where they support hypogeous, ectomycorrhizal fungi [19]. Like pines, this genus has been extensively planted outside its natural range in tropical countries, and like pines, its mycobiont community appears less diverse in these situations. *Acacia mangium* is another Australian species extensively used in tropical plantations. Apparently it too can form ectomycorrhizas, at least with *Pisolithus* sp. [20]. **Table 1** describes the list of ectomycorrhizal host taxa reported to contain at least one species forming this association.

Family	Genus
Aceraceae	*Acer*
Betulaceae	*Alnus, Betula, Carpinus, Corylus, Ostrya, Ostryopsis*
Bignoniaceae	*Jacaranda*
Caprifoliaceae	*Sambucus*
Casuarinaceae	*Casuarina, Allocasuarina*
Cistaceae	*Helianthemum, Cistus*
Compositae	*Lactuca*
Cyperaceae	*Kobresia*
Dipterocarpaceae	*Anisoptera, Balanocarpus, Cotylelobium, Dipterocarpus, Dryobalanops, Hopea, Monotes, Shorea, Valica*
Elaeagnaceae	*Shepherdia*
Epacridaceae	*Astroloma*
Ericaceae	*Arbutus, Arctostaphylos, Chimaphila, Gaultheria, Kalmia, Ledum, Leucothoe, Rhododendron, Vaccinium*
Euphorbiaceae	*Poranthera, Uapaca*
Fagaceae	*Castanea, Castanopsis, Fagus, Lithocarpus, Nothofagus, Pasania, Quercus, Trigonobalus*
Gentianaceae	*Bartonia*
Goodenaceae	*Brunonia, Goodenia*
Hammamelidaceae	*Parrotia*
Juglandaceae	*Carya, Juglans, Pterocarya*
Caesalpinoideae	*Afzelia, Aldina, Anthonota, Bauhinia, Brachystegia, Cassia, Eperua, Gilbertiodendron, Julbernardia, Monopetalanthus, Paramacrolobium, Swartzia*
Mimosoideae	*Acacia*
Papilionoideae	*Brachysema, Chorizema, Daviesia, Dillwynia, Eutaxia, Gompholobium, Hardenbergia, Jacksonia, Kennedya, Mirbelia, Oxylobium, Platylobium, Pultenaea, Robinia, Vicia, Viminaria*
Myricaceae	*Comptonia, Myrica*

Family	Genus
Myrtaceae	*Angophora, Callistemon, Campomanesia, Eucalyptus, Leptospermum, Melaleuca, Tristania*
Nyctaginaceae	*Neea, Torrubia, Pisonia*
Oleaceae	*Fraxinus*
Platanaceae	*Platanus*
Polygalaceae	*Comeosperma*
Polygonaceae	*Coccoloba, Polygonum*
Rhamnaceae	*Cryptandra, Pomaderris, Rhamnus, Spyridium, Trymalium*
Rosaceae	*Chaembatia, Cirocarpus, Crataegus, Dryas, Malus, Prunus, Pyrus, Rosa, Sorbus*
Salicaceae	*Populus, Salix*
Sapindaceae	*Allophylus, Nephelium*
Sapotaceae	*Glycoxylon*
Sterculiaceae	*Lasiopetalum, Thomasia*
Stylidiaceae	*Stylidium*
Thymeliaceae	*Pimelia*
Tiliaceae	*Tilia*
Ulmaceae	*Ulmus, Celtis*
Vitaceae	*Vitis*
Cupresseceae	*Cupressus, Juniperus*
Pinaceae	*Abies, Cathaya, Cedrus, Keteleeria, Larix, Picea, Pinus, Pseudolarix, Pseudotsuga, Tsuga*
Gnetaceae	*Gnetum*

Table 1.
Plant hosts reported to form ectomycorrhizal association.

4. Fungi forming ectomycorrhizal association

Approximately 5000–6000 fungal species form ectomyorrhizal association with the roots of trees. In recent years, more intensive mycological investigations in tropical forests and studies on hypogeous fungi associated with *Eucalyptus* vegetation in Australasia have unveiled numerous undescribed ectomycorrhizal species [19, 21, 22]. The use of molecular markers for direct mycobiont identification from ectomycorrhizas has significantly expanded the known taxa [23]. Furthermore, several fungal groups previously classified as saprotrophic, such as tomentelloid fungi, have been discovered to be ECM [24].

The contributing fungi are not evenly distributed over the kingdom *Fungi*. *Basidiomycota* is the major phylum in the kingdom that is involved about 95% in this association while members in *Ascomycota* play a minor role with approximately 5% ectomycorrhizal symbionts. Members in *Mucoromycota* only form exceptionally *ectomycorrhizal* associations [25].

Weiss et al. [23] revealed that the significance of heterobasidiomycetes within the *Sebacinaceae* as ectomycorrhizal symbionts has been underestimated. The members of this family have been strongly involved in forming this association [26]. As molecular studies on mycorrhizal symbionts increase, the taxonomic range of identified mycobionts is likely to broaden.

Division	Family	Genus	Species
Ascomycota	*Balsamiaceae*	*Balsamia*	*magnata, platyspora, vulgaris*
	Elaphomycetaceae	*Cenococcum*	*geophilum*
		Elaphomyces	*anthracinus, granulatus, muricatus, mutabilis, reticulatus, variegatus*
	Geneaceae	*Genabea*	*cerebriformis*
		Genea	*gardneri, harknessii, intermedia*
	Helvellaceae	*Hydnotrya*	*tulasnei*
	Pezizaceae	*Pachyphloeus*	*citrinus, ligericus, melanoxanthus*
		Paragalactinia	*succosella*
	Pyronemataceae	*Geopora*	*pinyonensis*
	Terfeziaceae	*Choiromyces*	*alveolatus, venosus*
	Tuberaceae	*Tuber*	*aestivum, borchii, brumale, californicum, excavatum, melanosporum, puberulum, rapaeodorum, rufum*
Basidiomycota	*Amanitaceae*	*Amanita*	*aspera, fulva, gemmata, inaurata, muscaria, pallidorosea, pantherina, phalloides, rubescens, solitaria, spissa, strobiliformis, subjunquillea, vaginata, verna, virosa*
	Astraeaceae	*Astraeus*	*hygrometricus, pteridus*
	Boletaceae	*Boletus*	*minatioolivaceus, pulverulentus, regius*
		Gyroporus	*castaneus, cyanescens*
		Phylloporus	*Rhodoxanthus*
		Pulveroboletus	*Ravenlii*
		Tylopilus	*chromapes, felleus, gracilis, porphyrosporus*
		Xerocomus	*armeniacus, badius, chrysenteron, rubellus, spadiceus, subtomentosus, truncatus*
	Cantharellaceae	*Cantharellus*	*cibarius, infundibuliformis, tubiformis*
	Clavariaceae	*Ramaria*	*aurea, botrytis, flava, formosa, mairei, subbotrytis*
	Corticiaceae	*Byssocorticium*	*Atrovirens*
		Byssoporia	*Sublutea*
		Piloderma	*byssinium, croceum, sulphureum*
	Cortinariaceae	*Cortinarius*	*acutus, anomalus, bicolor, biveluseverneus, hemitrichus, leucophanes, mucosus, multiformis, obtusus, phrygianus, saniosus*
		Dermocybe	*anthracina, cinnamomea, malicoria, palustris, phoenicea*
		Hebeloma	*crustuliniforme, cylindrosporum, hiemale, longicaudum, mesophaeum, minus, pumilum, sinapizans*
		Hymenogaster	*bulliardii, calosporus, citrinus, decorus, lilacinus, luteus, olivaceus, populetorum, tener, vulgaris*
	Inocybaceae	*Inocybe*	*asterospora, brunnea, cincinnata, curvipes, dulcamara, lacera, petiginosa*
		Inosperma	*adaequatum, bongardii*
		Mallocybe	*Terrigena*
		Pseudosperma	*Rimosum*
		Rozites	*Caperata*

Division	Family	Genus	Species
	Hydnaceae	*Dentinum*	*Repandum*
		Hydnellum	*Velutinum*
		Hydnum	*imbricatum, rufescens, scabrosum*
	Hygrophoraceae	*Hygrophorus*	*capreolarius, camarophyllus, chrysodon, discoideus, hypothejus, karstenii, marzulus, pudorinus*
	Hysterangiaceae	*Hysterangium*	*Membranaceum*
	Leucogastraceae	*Leucogaster*	*Nudus*
	Melanogastraceae	*Melanogaster*	*ambiguus, broomeianus, euryspermus, intermedius, tuberiformis, variegatus*
	Paxillaceae	*Paxillus*	*Involutus*
	Pisolithaceae	*Pisolithus*	*Tinctorius*
	Polyporaceae	*Albatrellus*	*Cristatus*
	Russulaceae	*Elasmomyces*	*Mattirolianus*
		Lactarius	*decipiens, fuliginosus, helvus, necator, piperatus, repraesentaneus, rufus, scrobiculatus, spinosulus, uvidus, vellereus, volemus*
		Russula	*aeruginea, albonigra, amoena, anthracina, cyanoxantha, densifolia, emetica, foetens, heterophylla, lutea, nigricans, ochroleuca, odorata, olivacea, paludosa, palumbina, parazurea, vesca, virescens, xerampelina*
		Zelleromyces	*Stephensii*
	Sclerodermataceae	*Scleroderma*	*bovista, cepa, citrinum, hypogaeum, laeve, polyrhizum, verrucosum*
	Strobilomycetaceae	*Boletellus*	*betula, chrysenteroides*
		Gautieria	*graveolens, mexicana, otthii*
		Strobilomyces	*floccopus*
	Thelephoraceae	*Thelephora*	*anthocephala, atrocitrina, penicillata, terrestris*
	Tricholomataceae	*Laccaria*	*amethystina, bicolor, laccata, montana, proxima*
		Tricholoma	*caligatum, columbetta, flavobrunneum, flavovirens, myomyces, saponaceum, sulphureum*
Mucoromycota	*Endogonaceae*	*Endogone*	*lactiflua*

Table 2.
Fungal partners known to form ectomycorrhizal association.

Certain ectomycorrhizal fungal genera are restricted to particular vegetation types. Members of *Inocybaceae* have been found forming association with *Pinaceae* in regions with a temperate climate [27]. *Cortinarius*, the most species-rich ectomycorrhizal genus, is prolific both in terms of species richness and sporocarp production in northern boreal regions [28], but is conspicuously absent from tropical regions [29]. By contrast, members of the *Russulaceae* show significant diversity in temperate and boreal regions and present a huge, largely undescribed, diversity in tropical regions where ECM hosts occur [22, 30, 31].

With the exception of some Tuber spp. [32], the knowledge of the ecology of ascomycete ECM fungi is very limited. Recent work [33, 34] suggests that the occurrence and importance of the *Helotiales* as ectomycorrhizal fungi may have been underestimated [35]. It demonstrated that an isolate from the *Hymenoscyphus ericae* aggregate [33] was capable of simultaneously forming ectomycorrhizal association on *Pinus sylvestris*. In recent years, the ectomycorrhiza of *Paragalactinia succosella* (=*Peziza succosella*) and *Geopora* spp. has also been reported with members of *Pinaceae* [36–39]. **Table 2** describes the list of ectomycorrhizal fungal taxa reported forming the ectomycorrhizal association.

5. Ectomycorrhiza in forestry

Since ectomycorrhizal fungi are known as integral to the ecosystem, and a huge diversity of these fungi form associations with a number of economically important tree species cultivated for timber, their potential importance in tree nutrition has been acknowledged since the earliest experiments [40]. Plantation forestry is on the rise globally due to the growing demand for timber and pulp, as well as efforts to boost carbon sequestration by expanding forested areas. Although seedlings that are outplanted or naturally regenerate can form ectomycorrhizal associations from any naturally present mycorrhizal inoculum, the evident benefits of nursery stock being mycorrhizal prior to outplanting are apparent.

Many commercial practices, especially those aimed at enhancing hygiene in tree nurseries, are detrimental to the growth of most ectomycorrhizal fungi species, except for a few ruderal ones. As a result, specific techniques have been devised to facilitate the colonization of plants by selected fungi before outplanting. The application of these techniques has facilitated superior performance in tree crops in many parts of the world, particularly those which lack natural local sources of inoculum [41]. Grove and Le Tacon [42] provide a comprehensive outline of the imperatives for developing successful strategies to maximize the benefits of ectomycorrhizas in forestry. Meanwhile, there has been an increasing interest in the potential of harvesting edible fruiting bodies of ECM fungi, which have been utilized as commercial inoculants to enhance both diet and revenue. Inoculation can provide benefits at two key stages of timber production systems: in the nursery phase and post-outplanting in the field.

A significant portion of experimental research has concentrated on the benefits associated with producing well-developed seedlings that, along with their fungal symbionts, can thrive once transplanted into the field. While this focus primarily aims to ensure successful establishment, there are also direct cost savings realized through the accelerated throughput resulting from rapid growth in nurseries.

Furthermore, the utilization of inoculated seedlings has provided valuable insights. It has been observed that the responses to ectomycorrhizal colonization are particularly pronounced under challenging conditions, such as drought, metal contamination, and pathogen exposure. These observations have prompted an in-depth analysis of the functional mechanisms underlying the beneficial effects of ectomycorrhizal fungi, shedding light on their role in ameliorating adverse environmental factors.

Beyond the extensively studied effects on plant nutrition and growth, significant progress has been achieved in comprehending the roles of symbiotic relationships in imparting resistance to various stresses. These stresses, while occurring naturally in ecosystems, are often exacerbated locally due to prior land-use practices or the afforestation process itself. Discrepancies in performance between nursery and field conditions following inoculation may indeed arise from differential impacts of soil and environmental conditions on the specific mycorrhizal partnerships established through inoculation.

Forest productivity faces threats from global change factors such as increased inputs of nitrogen and sulfur directly into soils, leading to simultaneous decreases in pH, which can negatively affect both symbiotic partners. These effects are believed to contribute to forest decline syndromes witnessed in various parts of the world. Additionally, there are indications that the availability of base cations may eventually constrain forest productivity, emphasizing the potential benefits of ectomycorrhizal associations in enhancing nutrient uptake under such circumstances.

Acknowledgements

Sincere thanks to my students; Memoona Azeem, Wasiqa Arshad, and Mehboobullah Khan who helped me in formatting the text.

Conflict of interest

The author declares no conflict of interest.

Author details

Sana Jabeen
Department of Botany, Division of Science and Technology, University of Education, Lahore, Punjab, Pakistan

*Address all correspondence to: sanajabeenue@gmail.com; sanajabeen@ue.edu.pk

References

[1] Colpaert JV. Heavy metal pollution and genetic adaptations in ectomycorrhizal fungi. In: British Mycological Society Symposia Series, 27. Cambridge, Massachusetts, United States: Academic Press; 2008. pp. 157-173

[2] Finlay RD, Lindahl BD, Taylor AF. Responses of mycorrhizal fungi to stress. In: British Mycological Society Symposia Series, 27. Cambridge, Massachusetts, United States: Academic Press; 2008. pp. 201-219

[3] Smith SE, Read DJ. Mycorrhizal Symbiosis. 3rd ed. London: Academic Press Elsevier; 2008

[4] Fitter AH, Moyersoen B. Evolutionary trends in root-microbe symbioses. Philosophical Transactions of the Royal Society of London Series B: Biological Sciences. 1996;**351**:1367-1375

[5] Hibbett DS, Matheny PB. The relative ages of ectomycorrhizal mushrooms and their plant hosts estimated using Bayesian relaxed molecular clock analyses. BMC Biology. 2009;**7**:13

[6] Alexander IJ. Ectomycorrhizas—Out of Africa? New Phytologist. 2006;**172**:589-591

[7] Lepage BA, Currah RS, Stockey RA, Rothwell GW. Fossil ectomycorrhiza from the middle Eocene. American Journal of Botany. 1997;**84**(3):410-412

[8] Meyer FH. Distribution of ectomycorrhizae in native and man-made forests. In: Marks GC, Kozlowski TT, editors. Ectomycorrhizae. New York: Academic Press; 1973. pp. 79-105

[9] Read DJ. The ties that bind. Nature. 1997;**388**:517-518

[10] Massicotte HB, Melville LH, Peterson RL, Luoma DL. Anatomical aspects of field ectomycorrhizas on *Polygonum viviparum* (*Polygonaceae*) and *Kobresia bellardii* (*Cyperaceae*). Mycorrhiza. 1998;7:287-292

[11] Read DJ, Perez-Moreno J. Mycorrhizas and nutrient cycling in ecosystems—A journey towards relevance? New Phytologist. 2003;**2003**(157):465-492

[12] Taylor AF, Alexander IAN. The ectomycorrhizal symbiosis: Life in the real world. Mycologist. 2005;**19**(3):102-112

[13] Ducousso M, Bena G, Bourgeois C, Buyck B, Eyssartier G, Vincelette M, et al. The last common ancestor of *Sarcolaenaceae* and Asian dipterocarp trees was ectomycorrhizal before the India-Madagascar separation, about 88 million years ago. Molecular Ecology. 2004;**13**:231-236

[14] Newbery DM, Alexander IJ, Thomas DW, Gartlan JS. Ectomycorrhizal rain-forest legumes and soil phosphorus in Korup National Park, Cameroon. New Phytologist. 1988;**109**(4):433-450

[15] Alexander IJ. Systematics and ecology of ectomycorrhizal legumes. Advances in Legume Biology, Monographs in Systematic Botany. 1989;**29**:607-624

[16] Munyanziza E, Oldeman RAA. Miombo trees: Ecological strategies, silviculture and management. Ambio. 1996;7:454-458

[17] Verbeken A, Buyck B. Diversity and ecology of tropical ectomycorrhizal fungi in Africa. In: Tropical Mycology: Vol. 1,

Macromycetes. Wallingford, UK: CABI Publishing; 2002. pp. 11-24

[18] Read DJ. The mycorrhizal status of *Pinus*. In: Richardson DM, editor. Ecology & Biogeography of *Pinus*. Cambridge, United Kingdom: Cambridge University Press; 1998. pp. 324-340

[19] Claridge AW. Ecological role of hypogeous ectomycorrhizal fungi in Australia forests and woodlands. Plant and Soil. 2002;**244**:291-305

[20] Duponnois R, Ba AM. Growth stimulation of *Acacia mangium* Willd by *Pisolithus* sp. in some Senegalese soils. Forest Ecology and Management. 1999;**119**:209-215

[21] Haug I, Weiss M, Homeier J, Oberwinkler F, Kottke I. *Russulaceae* and *Thelephoraceae* form ectomycorrhizas with members of the *Nyctaginaceae* (*Caryophyllales*) in the tropical mountain rain forest of southern Ecuador. New Phytologist. 2005;**165**:923-936

[22] Buyck B, Thoen D, Watling R. Ectomycorrhizal fungi of the Guinea Congo region. Proceedings of the Royal Society of Edinburgh, Section B: Biological Sciences. 1996;**104**:313-333

[23] Weiss M, Selosse MA, Rexer KH, Urban A, Oberwinkler F. Sebacinales: A hitherto overlooked cosm of heterobasidiomycetes with a broad mycorrhizal potential. Mycological Research. 2004;**108**:1003-1010

[24] Kõljalg U, Dahlberg A, Taylor AFS, Larsson E, Hallenberg N, Stenlid J, et al. Diversity and abundance of resupinate thelephoroid fungi as ectomycorrhizal symbionts in Swedish boreal forests. Molecular Ecology. 2000;**9**:1985-1996

[25] Molina R, Massicotte H, Trappe JM. Specificity phenomenon in mycorrhizal symbiosis: Community ecological consequences and practical implications. In: Allen MF, editor. Mycorrhizal Functioning. London, UK: Chapman and Hall; 1992. pp. 357-423

[26] Urban A, Weiss M, Bauer R. Ectomycorrhizae involving sebacinoid fungi and their analogs. Journal of the Linnaean Society of London, Botany. 2003;**13**:31-42

[27] Jabeen S. Ectomycorrhizal Fungal Communities Associated with Himalayan Cedar from Pakistan [PhD Thesis]. Lahore, Pakistan: Department of Botany, University of Punjab; 2016

[28] Brandrud TE. The effects of experimental nitrogen addition on the ectomycorrhizal fungus flora in an oligotrophic spruce forest at Gårdjön, Sweden. Forest Ecology and Management. 1995;**71**:111-122

[29] Peintner U, Moser MM, Thomas KA, Manimohan P. First records of ectomycorrhizal *Cortinarius* species (*Agaricales*, *basidiomycetes*) from tropical India and their phylogenetic position based on rDNA ITS sequences. Mycological Research. 2003;**107**:485-494

[30] Henkel TW, Terborgh J, Vilgalys RJ. Ectomycorrhizal fungi and their leguminous hosts in the Pakaraima mountains of Guyana. Mycological Research. 2002;**106**:515-531

[31] Lee SS, Watling R, Turnbull E. Diversity of putative ectomycorrhizal fungi in Pasoh forest reserve. In: Okuda T, Manokaran N, Matsumoto Y, Niiyama K, Thomas SC, Ashton PS, editors. Pasoh: Ecology of a Lowland Rain Forest in Southeast Asia PS. Tokyo: Springer; 2003. pp. 149-159

[32] Murat C, Díez J, Luis P, Delaruelle C, Dupre C, Chevalier G,

et al. Polymorphism at the ribosomal DNA ITS and ITS relation to postglacial re-colonization routes of the Perigord truffle *tuber melanosporum*. New Phytologist. 2004;**164**:401-411

[33] Vrålstad T, Fossheim T, Schumacher T. *Piceirhiza bicolorata*—The ectomycorrhizal expression of the *Hymenoscyphus ericae* aggregate? New Phytologist. 2000;**145**:549-563

[34] Vrålstad T, Schumacher T, Taylor AFS. Mycorrhizal synthesis between fungal strains of the *Hymenoscyphus ericae* aggregate and potential ectomycorrhizal and ericoid hosts. New Phytologist. 2002;**153**:143-152

[35] Villarreal-Ruiz L, Anderson IC, Alexander IJ. Interaction between an isolate from the *Hymenoscyphus ericae* aggregate and roots of *Pinus* and *Vaccinium*. New Phytologist. 2004;**164**:183-192

[36] Jabeen S, Ashraf T, Khalid AN. *Peziza succosella* and its ectomycorrhiza associated with *Cedrus deodara* from Himalayan moist temperate forests of Pakistan. Mycotaxon. 2015;**130**(2):455-464

[37] Jabeen S, Asghar HS, Niazi AR, Khalid AN. *Geopora asiatica* sp. nov. from Pakistan. Nova Hedwigia. 2022;**114**(3-4):389-399

[38] Long D, Liu J, Han Q, Wang X, Huang J. Ectomycorrhizal fungal communities associated with *Populus simonii* and *Pinus tabuliformis* in the hilly-gully region of the loess plateau, China. Scientific Reports. 2016;**6**(1):24336

[39] Guo M, Gao G, Ding G, Zhang Y. Drivers of ectomycorrhizal fungal community structure associated with *Pinus sylvestris* var. *Mongolica* differ at regional vs. local spatial scales in northern China. Forests. 2020;**11**(3):323

[40] Frank AB. Über die auf Wurzelsymbiose beruhende Ernährung gewisser Bäume durch unterirdische Pilze. Berichte der Deutschen Botanischen Gesellschaft. 1885;**3**:128-145

[41] Marx DH, Marrs LF, Cordell CE. Practical use of mycorrhizal fungal technology in forestry, reclamation, arboriculture, agriculture and horticulture. Dendrobiology. 2002;**47**:29-42

[42] Grove TS, Le Tacon F. Mycorrhiza in plantation forestry. Advances in Plant Pathology. 1993;**9**:191-228